The Mind of Everyday

Petraq Papajorgji • Howard Moskowitz

The Mind of Everyday

Combining Individual and Artificial Intelligence

Springer

Petraq Papajorgji
Moore Albania SH.P.K.
Tirana, Albania

Howard Moskowitz
Tactical Data Group
Stafford, VA, USA

ISBN 978-3-031-78077-6 ISBN 978-3-031-78078-3 (eBook)
https://doi.org/10.1007/978-3-031-78078-3

This Springer imprint is published by the registered company Springer Nature Switzerland AG
The registered company address is: Gewerbestrasse 11, 6330 Cham, Switzerland

Foreword

From the early days of artificial intelligence (AI) to the present, the question of how well AI-based systems can emulate the human brain has been a focal point of lively and ongoing discussions in academic and engineering circles. This dynamic discourse, which addresses both philosophical and engineering perspectives, continues to shape the field of AI. In 1950, Turing published his paper "Computing Machinery and Intelligence," which discussed the computer's ability to exhibit intelligent behavior equivalent to or indistinguishable from a human's (TURING, 1950). Turing posed the question, "Can machines think?" and described a process to operationalize this question without getting involved in the philosophical debate over what it means to "think." This process was called "The Turing Test."

The Turing Test, a concept that has echoed through the halls of AI research, consisted of a human judge engaging in a natural language conversation with both a human and a machine located in separate environments without knowing which was which. If the judge cannot reliably distinguish the machine from the human, the machine is considered to have passed the test, demonstrating a form of artificial intelligence. The Turing Test has been influential in discussions of AI, though it has also been criticized for its focus on imitation rather than proper understanding or consciousness. One of the most vocal critics was Lady Lovelace, alias Ada Bayron, known as the world's first programmer. According to Ada Byron, machines can only do what they are programmed to do and cannot originate anything new. Of course, Lady Lovelace lived a century before Turing, so the words "vocal critic" must be taken as a metaphor for their very real disagreement!

Defending his conjecture, Turing reasoned that there is no good reason to deny the possibility that machines could, 1 day, eventually think in a manner not discernible from that of humans. He suggested that the question of whether machines can think should be replaced with the more operational and practical question regarding whether or not machines can behave in ways that are indistinguishable from humans, a shift in perspective which has profound implications for the practical applications of AI.

In mid-2024, the world is witnessing something new, never seen before. A wave of new technologies, mainly composed of AI and synthetic biology and helped by quantum computing, is starting to affect all aspects of our lives. Suleyman and Bhaskar (2023) present the economic, social, and political impacts of this new wave with exciting facts, arresting arguments, and compelling observations.

Although the new technological wave is all-encompassing and omnipotent, the issue of comparing human and technology performance is still valid and needs to be addressed. This issue is the main target of this book.

To address these issues, the authors came up with the notion of investigating the world of the mind using the inspiration provided by psychophysics coupled with the possibilities offered by artificial intelligence. The basic idea treated in this book is simple: find ways to measure thought, or at least ideas, and then compare ideas generated by people to ideas generated by AI. In other words, are the ideas generated by AI any "better" than the ideas generated by people when the evaluator is a human being? The direct result of this challenge is the backbone of "The Mind of Everyday: Combining Individual and Artificial Intelligence."

The book has presented the topic in a structure which guides the reader step by step with a sensitivity to the technology, but also to the human being living with the technology and confronting the future.

Chapter 1 begins by laying out the "raison d'etre" of the authors' efforts, introducing the emerging science of Mind Genomics. Mind Genomics has roots in several fields, the most significant of which is psychophysics. A psychophysicist is a measurement specialist who uses the mind as a measuring tool. Later in this chapter, the authors introduce AI, which has enormous power to perform vast amounts of data. A marriage between Mind Genomics and AI seems rational and even unavoidable.

Ordinary linear regression is often used, and it is one of the building blocks of the Mind Genomics shell, which offers these functionalities as a service. Chapter 2 reviews the topic linear regression simply and intuitively, designed for social and political audiences, who may not be involved with these technologies. Thus, students may not need to search for other statistical knowledge sources, but rather use this chapter.

A well-designed research question is a crucial step in the research process. It requires time, expertise, and consideration to ensure the question is well-structured, meaningful, and feasible. Therefore, the authors dedicate Chap. 3 to clarify and explain the differences between hypothesis-based versus hypothesis-free research. This chapter includes advanced and intensive debates within the research community about the best approach. The advantages and disadvantages of each method are presented, and advice is given on when one approach is more suitable than the other, providing the researcher with the full spectrum of issues to consider when selecting the research method.

Since the early days of recorded civilization, philosophers have wondered how they know, what is considered knowledge, and how knowledge increases. Chapter 4 talks about the two main approaches to understand and explain everyday life phenomena, the qualitative and quantitative approaches, respectively. For quite some time, there has been a vivid debate in the scientific community regarding the relative virtues of the quantitative and qualitative "methods" and worldviews, followed by questions such as which methods are "better," or more reasonably, what are the conditions which favor using one of the two classes of measuring, and why.

Chapter 5 deals with AI and provides definitions, a history of how it started, and the latest developments in this complex, breathtaking technology that is set to drastically change how we live today. The authors provide ample examples to demonstrate how AI works without using a complex mathematical approach to make it easier for the social and political audience. As they offer multiple examples of the wonders AI provides, the authors are both enthusiastic and cautious in recommending the use of AI.

Chapter 6 introduces Mind Genomics, the aforementioned emerging science which focuses on understanding daily experience and, more specifically, the decisions we make about the everyday. Mind Genomics traces its history back to several areas, the most important being psychophysics. A psychophysicist focuses on measurement, with the measuring instrument being the mind. Traditional psychophysics concentrated on what we might call "external" stimuli, whether the physical sound level of a bell, the concentration of milk in a cup of coffee, and so forth. One could recognize the physical measures. They are obvious and standard. But what about the perception? What if one doubled the amount of milk in the coffee—would the coffee taste twice as milky? The idea of a "meter" is one of the foundations of this book. The authors are interested in measuring the magnitudes in the mind, something which may sound strange but is really the daily "bread and butter" of psychophysics.

The advent of AI impacted the way Mind Genomics uses the "meter." With its immense calculating power, AI gave Mind Genomics another dimension. AI-powered Mind Genomics empowers analytics and provides different development scenarios that humans cannot offer.

The last four chapters are applications of this research philosophy. These essential topics cover education, family stress, environmental policies, and governmental policies, respectively. The study's overall goal is to determine whether there is a difference between human-driven and AI-driven studies. How effectively does AI "think" concepts compared to humans, or perhaps even surpass them?

Like the Turing Test, humans judged the results of human- and AI-based tests. Unlike the Turing Test, however, human and AI-based performances are evaluated using the same approach. The authors make it possible to compare general patterns emerging from the two studies on the same topic to evaluate the results obtained by human and AI-based approaches. The authors suggest several ways to compare the patterns of responses generated by the two experiments, ranging from the patterns of ratings, the relations between the presence/absence of elements versus the response to the vignette (modeling), the interpretation of strong performing elements (viz., interpretation and AI-based summarization), and finally the division of the respondents into mindsets by clustering (Mucherino et al. 2009) and the nature of these clusters.

The authors use the Index of Divergent Thought (IDT) (Thakral et al. 2021; Madore et al. 2016; Giancola et al. 2023) as a paramount tool in this study, enabling the evaluation of each approach's performance. The IDT is introduced as a computational formula to assess how successfully the Mind Genomics initiative gathered information about people's thoughts, and it is introduced as a computational formula to demonstrate how well the ideas worked.

Experience showed that it was becoming apparent that with AI it was very easy to get high IDT values, corresponding to the fact that the elements or messages studies appealed to everyone. Yet, the goal is to create messages which appeal to some people and not to others. The authors ran simulations to show that IDT values around 70 corresponded to strong-performing elements in some mindsets and not in others, a finding that is more meaningful than having strong-performing elements in all mindsets. It was the human-generated elements which tended to have IDT values around 70, suggesting that AI may optimize, but is still lacking the ability to understand and capitalize on the subtle differences among mindsets. The ability to sense the nuances may lead to lower performances overall in comparative tests, but that same ability to sense nuances and thus suboptimize the messages does not yet seem to be in the purview of AI as of this writing (summer, 2024).

The book offers a solid and holistic approach to evaluate AI performance vis-à-vis humans. The authors provide the entire set of knowledge and tools to allow for the analysis of this complex issue and the drawing of a final result. Although care is taken to use simple mathematical presentations, the design of the problem, the presentation of philosophical approaches, and the mathematical tools used to depict results are excellent. The method used is simple and conforms to the **technology-as-a-service** approach.

This book offers many valuable insights to students and specialists in political and social science. It also offers a number of complex technologies, such as linear regression, data mining, and clustering technologies, under a web-based, easy-to-use application. In addition, this book represents one of the few efforts that use advanced AI-powered quantitative technologies to address social and political issues.

Considering the relevance of the topics analyzed, the technologies used, and the results obtained, the title *The Mind of Everyday: Combining Individual and Artificial Intelligence* deserves to be on the books for serious social and political scientists.

Distinguished Professor, Georgia Institute of Technology Ian Akyildiz,
Atlanta, GA, USA

References

Giancola, M., D'Amico, S., & Palmiero, M. (2023). Working Memory and Divergent Thinking: The Moderating Role of Field-Dependent-Independent Cognitive Style in Adolescence. *Behavioral Sciences*, *13*(5), 397. https://doi.org/10.3390/bs13050397

Madore, K. P., Jing, H. G., & Schacter, D. L. (2016). Divergent creative thinking in young and older adults: Extending the effects of an episodic specificity induction. *Memory & Cognition*, *44*(6), 974–988. https://doi.org/10.3758/s13421-016-0605-z

Mucherino, A., Papajorgji, P., & Pardalos, P. (2009). Clustering by k-means. In *Data Mining in Agriculture* (pp. 47–82).

Suleyman, M., & Bhaskar, M. (2023). *The Coming Wave: Technology, Power, and the Twenty-first Century's Greatest Dilemma*. Crown.

Thakral, P. P., Yang, A. C., Addis, D. R., & Schacter, D. L. (2021). Divergent thinking and constructing future events: dissociating old from new ideas. *Memory*, *29*(6), 729–743. https://doi.org/10.1080/09658211.2021.1940205

TURING, A. M. (1950). I.—COMPUTING MACHINERY AND INTELLIGENCE. *Mind*, *LIX*(236), 433–460. https://doi.org/10.1093/mind/LIX.236.433

Preface

Whether we like it or not, the advent of artificial intelligence (AI) is associated with relevant changes in the way we live today. These changes are profound and multifaceted and will affect all aspects of our lives: material, cultural, and even spiritual.

AI-based translators make it easy and possible to speak to anyone in most parts of the world without needing a translator. AI-based document management systems allow for translating documents into many other languages, summarizing documents quickly, automated data extraction and classification, and optical character recognition that allows scanned documents to be edited and searchable.

Not too long ago, Siri, Google Assistant, and Alexa were *technological wonders*. Nowadays, almost every significant system has AI-based customer services helping people navigate through the labyrinths of large information systems.

AI revolutionizes healthcare, especially patient care, treatment planning, and diagnosis. AI-based systems can evaluate medical data, including test results and scans, helping physicians identify illnesses early on with admirable accuracy. An exemplary contribution of artificial intelligence in healthcare is Inflection AI's Pi, an AI assistant with emotional intelligence specifically developed to participate in sympathetic, human-like interactions, primarily intended to support elderly individuals.

AI has the potential to significantly enhance daily decision-making by providing tailored, data-dependent insights, automating repetitive activities, and aiding in intricate analysis.

Thus, AI-based systems will be around us, helping and advising us in our everyday tasks. However, how well do these systems perform in their quest to "replace" the human brain? Do ideas generated by AI-based systems perform better than ideas generated by humans?

To address this issue, a "meter," a procedure for comparing the outcomes obtained by artificial intelligence systems with those obtained by human systems, is necessary. This "meter" is Mind Genomics, a new emerging science with a focus on understanding how people make everyday decisions.

The German polymath Gustav Theodor Fechner initiated the idea of using a "meter" to measure feelings in the middle of the 1800s. Fechner was looking for the inner measuring instrument of feeling and broached the topic of "measuring private sensory experience." He was attempting to find the instrument that would measure sensation. S.S. (Smitty) Stevens, a professor of psychology at Harvard University, advanced the idea of measuring sensation to a higher level. Stevens conducted tests employing this measurement in a new and more direct method, which allowed him to elevate this theory.

Stevens proposed that the individual/observer could be instructed to assign a number to match the degree of appreciating something. For example, the observer may provide a number matching the tea's perceived sweetness or strength of the coffee flavor. Stevens demonstrated that it is possible to match the size of the numbers of the probability estimated by individuals. Thus, this is the beginning of the concept that one can measure the sensory reaction, an experience that takes place inside oneself.

The more research that was carried out, the more obvious it became that these studies could be conducted with relative ease. All that would need to be done would be to create "test stimuli" by varying the physical stimuli in an orderly manner, present them in a random order, and then obtain the respondent's ratings of perceived magnate. This research approach was called "Outer Psychophysics."

The required "meter" to measure the perceptions of our everyday world is Mind Genomics. Mind Genomics delves into the science of how we, as humans, organize and respond to information, unlocking the hidden patterns of thought, emotion, and behavior that shape our perceptions and actions in our everyday life.

To evaluate whether there are differences in the results obtained by AI-based systems and human-based experiments based on "Outer Psychophysics" were designed and implemented. Thus, four experiments in crucial areas of our lives are selected. These essential areas cover education, family stress, and environmental and governmental policies. For each area, a Mind Genomic experiment was designed, defining four topics (silos). Then, the elements contained in each of the silos were generated twice, once by humans and then by an AI-based system. Evaluating both systems' performance requires comparing the results obtained by both experiments.

Comparing the obtained results is not straightforward; it requires discovering inner relationships among the data. The last four chapters of this book offer a wide range of tools to understand patterns of responses, such as the patterns of ratings, the relationships between elements' presence or absence and response to the vignette (modeling), the interpretation of high-performing elements (i.e., interpretation and AI-based summarization), and, lastly, the clustering of respondents into mindsets and the makeup of these clusters.

Each study case presents different combinations of patterns and solutions to address the problem. Readers will find a plethora of methods that can be used in future studies. The analysis of the study cases is disciplined, rich in scientific tools, and presented with methods, so it is easy to follow the same path when necessary. Through a blend of theoretical exploration, case studies, and practical applications, we aim to inspire readers with the potential of the fusion of Mind Genomics and AI to empower innovation, enhance user experiences, and bring forth a more nuanced understanding of human thought.

Although explaining AI and how to use it to address scientific issues is a complex phenomenon, extra care is taken to make the presentation of this book effortless and pleasant for audiences of social and political sciences who are not savvy in mathematics.

We hope readers will not only enjoy reading this book but also feel inspired to apply the presented approach to address other and even more complex issues in their work. Whether the reader is a student, a researcher in the field of social and political sciences, or simply curious about the evolving landscape of science and technology, we hope this book will serve as an invitation to explore how we can harmonize human and machine intelligence to better navigate the complex world around us.

This book is a tool for empowerment, and we encourage readers to explore its potential in their respective fields.

Tirana, Albania
New York, NY, USA
July 2024

Petraq Papajorgji
Howard Moskowitz

The original version of the book has been revised. A correction to this book can be found at https://doi.org/10.1007/978-3-031-78078-3_11

Contents

Chapter 1
The Purpose of This Work

1.1 Introduction

Mind Genomics is an emerging science, which focuses on understanding daily experiences and, more specifically, the ordinary decisions that we make every day, often doing so automatically. Mind Genomics is a young field, still evolving.

Mind Genomics traces its history back to a number of areas, the most important being psychophysics. A psychophysicist focuses on measurement, with the measuring instrument being the mind. Traditional psychophysics concentrated on what we might call "external" stimuli, whether those be the physical sound level of a bell, the concentration of milk in a cup of coffee, and so forth. One could recognize the physical measure; they were obvious and standard. But what about the perception? What if one doubled the amount of milk in the coffee—would the coffee taste twice as milky?

The idea of a "meter" is one of the foundations of this book. We are interested in measuring the magnitudes in the mind, something that may sound strange, but is really the daily "bread and butter" of this area called psychophysics. But the quest is more. A book devoted to the original psychophysics has already been published, indeed by the doyen of psychophysics, S. S. Stevens of Harvard University. That book is "Psychophysics: An introduction to its perceptual, neural and social prospects," published in 1975, 2 years after Stevens' untimely passing from heart failure at age 67.

So, if Stevens and his many colleagues worldwide already dealt with the problem, why this book a half century later? What does this book add to what we know about the milkiness of a coffee, the loudness of a sound, the brightness of a light? In the colloquial language "*been there, done that*."

The subject of this book is the "Inner Psychophysics," the measurement of ideas, the measurement of feelings. We deal here with a new way of measuring private experience, not so much the sensory experience hitherto the topic of research, but

P. Papajorgji, H. Moskowitz, *The Mind of Everyday*,
https://doi.org/10.1007/978-3-031-78078-3_1

the topic of how strongly we feel about ideas, often ideas to which we pay little attention in our daily lives.

During the middle of the 1800s, the German polymath Gustav Theodor Fechner, who was looking for the inner measuring instrument of feeling, broached the topic of "measuring private sensory experience." Fechner was attempting to find the instrument that would measure sensation. According to Heidelberger (1994) and Ortlieb et al. (2020), a search was conducted to locate the internal measuring device of feeling. The fact that a person was experiencing something, like enjoying the sweetness of a cup of tea, was not sufficient evidence to prove that they were conscious of what they were experiencing. "*What was the perceived strength of sweetness*?" was the question needing to be addressed and answered. One of the ways in which Fechner addressed the topic was conducting a philosophical investigation that dealt with the concept of measuring "private sensory experience." It is generally agreed that he was the first person to state that sensations could be measured scientifically (Bruch & Feinberg, 2017).

This concept was taken to a higher level by S.S. (Smitty) Stevens, the aforementioned professor of psychology at Harvard University, and doctoral advisor to author Moskowitz of this volume. Stevens conducted tests employing this measurement in a new and more direct method, which allowed him to take this theory to a higher elevation (Stevens 1975). It was proposed by Stevens that the individual, the "observer" or "O" from German word "Beobachter," could be instructed to assign a number to match the degree of appreciating something. As an example, the "observer" may provide a number matching the perceived sweetness of the tea or perceived strength of the coffee flavor. In the case of individuals, but particularly with groups of people, Stevens demonstrated that it is possible to match the size of the numbers of the probability estimated by individuals. What was being addressed was the concept that one was measuring the sensory reaction, which is an experience that takes place inside oneself.

As further research was carried out, it became increasingly apparent that one could do these studies quite easily, simply by varying the physical stimuli in an orderly fashion to create "test stimuli," present these stimuli in random order, and then get the ratings of perceived magnate as assigned by the respondent. This research approach was called the "Outer Psychophysics."

Thus, we finish the first half of the inspiration: the contribution of the old discipline of psychophysics to measuring the private perception of well-defined physical stimuli. The next half of the inspiration was needed, viz., measuring the private world of ideas.

In the chronicles of technological history, the introduction of CHATGPT (Generative Pre-Trained Transformer), developed by OpenAI in November 2023, turned out to be something that can be only called "one of a kind," unique, sui generis. In a very short time, the number of individuals who indicated an interest in experimenting with the new technology reached millions, something which never happened before in the history of technology.

Over the course of a very short period of time, it became abundantly evident that the technological environment was on the cusp of experiencing a significant upheaval. At this very moment of writing, we are witnessing increasing enthusiasm and buzz around the prospects possible to everyone thanks to breakthroughs in generative artificial intelligence.

Another point that should be emphasized is the rapid rate at which artificial intelligence is being developed. It is estimated that about once every 2 to 3 months, a significant news article emerges, once again capturing the attention of the ever-increasing interest public, once again putting artificial intelligence (AI) into the limelight.

We came up with the notion to investigate the world of the mind using the inspiration provided by psychophysics, coupled with the possibilities offered by artificial intelligence. The basic idea was to measure thought, or at least ideas, comparing ideas generated by people to ideas generated by AI. In other words, are the ideas generated by AI any "better" than the ideas generated by people, when the evaluator is a human being. The direct result of this challenge is "The Mind of Everyday: Combining Individual and Artificial Intelligence." The objective is to introduce students and professionals working in the field of social sciences to a new way of thinking about issues (Mind Genomics). Rather than simply presenting ideas and possibilities, we do experiments to show substantive results.

1.2 Artificial Intelligence and the World of Social Science

The introduction of artificial intelligence (AI) into the domains of political and social sciences signals a new era of intellectual discourse. AI provides new ways to approach practical as well as theoretical issues, addressing problems with novel solutions, and perhaps even with intriguing albeit perplexing ideas. Social sciences may significantly advance with the new tools provided by artificial intelligence. There are no right/wrong answers, so the contribution will be analytical, ways to manage data, ways to produce ideas, ways to suggest "next steps." The tools or "instruments" provided by artificial intelligence will allow professionals, government, citizens alike to deal with the often challenging, daunting, even frightening array of complex social phenomena, such as election patterns, shifts in public opinion, the rise of tribal mentalities among groups of people, and so forth.

The ability to review, assess, summarize, and make recommendations, all with vast volumes of data, catapults artificial intelligence almost immediately into a potent player of the world of knowledge. The ability of artificial intelligence to filter through vast datasets, such as posts on social networking platforms, public records, and news items, and provide a more comprehensive perspective than more conventional methods of analysis is simply breathtaking to those who want to enumerate the possibilities available right now, as of this writing (Summer, 2024). The weapons of yesteryear may soon fade into memories when we realize that artificial intelligence, condensed and yet malleable knowledge, constitutes a more powerful weapon than the ones that have traditionally been employed in warfare.

One of the most rapidly expanding topics in the fields of social science is the use of artificial intelligence in practical settings. This subject encompasses a broad variety of applications and is enjoying rapid growth. Applications built on artificial intelligence provide opportunities for analysis, prediction, and enhanced policymaking never before known.

All is not "rosy," however. With the new and powerful comes the fear of misuse, the fear of incompetence when given a powerful weapon and little training, little experience. The aspiring user of artificial intelligence should be counseled to investigate the risks that are associated with it, not just rejoice in its promise and capabilities. When it comes to the domains of political and social sciences, it is not out of the question for artificial intelligence both to contribute to and to challenge the conventional wisdom now prevalent. The challenge is to develop new theories, but do so in a disciplined, fair, and ethical manner, one which leads to a future unbroken path of intellectual and then practical advances.

AI has changed the landscape of scientific research drastically. One may ask the next "logical question," appropriate for those who are deeply involved in the "how things are done," the so-called "methodologists." Note that we use the word "methodology" in the correct manner, as the science of method.

Returning to AI, one may ask the very simple "methods" question, namely, "where does AI fit?" Is AI part of the qualitative arsenal of tools or, instead, part of the quantitative set of tools? Qualitative research deals with ideas; quantitative research deals with measuring.

When it comes to qualitative research, artificial intelligence may also contribute, mostly via the processing and analysis of non-numerical data such as text, photographs, or videos, identifying patterns, telling stories, working with "stuff" that is not typically measured.

When it comes to quantitative methods, with numbers, we deal with machine learning and deep learning, systems that identify pattern. Machine learning and deep learning often are tasked to evaluate massive amounts of numerical data, first to discover patterns, and second to generate predictions. To provide just one example, artificial intelligence algorithms are able to evaluate vast volumes of weather data in order to establish correlations, carry out classifications, or forecast weather events based on quantitative inputs, respectively.

The algorithms used to analyze text data, classify topics, comprehend sentiment, and recognize patterns in a manner analogous to qualitative analysis fall under the category pf artificial intelligence known as Natural Language Processing (NLP). NLP methods handle larger volumes of data than human researchers are able to manually manage, allowing NLP to provide meaningful insights from unstructured data that might only be apparent with the assistance of computational assistance.

Artificial intelligence is playing an increasingly important role in mixed-methods research, methods that combine qualitative and quantitative procedures. AI bridges the gap between these two worlds by providing tools, which quantitatively analyze qualitative data. Examples of this bridge are sentiment analysis and thematic analysis. AI generates qualitative insights from the quantitative data, talking about "whys," not just "whats."

1.3 The World of the Everyday: Understanding People by AI-Driven Mind Genomics

The book we present "The Mind of Everyday: Combining Individual and Artificial Intelligence," explains quantitative research methods and artificial intelligence in a way understandable to readers who are not technically savvy. The book focuses on understanding how people make decisions about topics of political science using the tools of quantitative research coupled with artificial intelligence. In a sense the book is a pioneering effort, presenting old and new together, in the attempt to secure more wisdom about "*homo politicus*," man (man and woman) in a society of law.

The quantitative structure undergirding this book is the emerging science of Mind Genomics, the science of "inner psychophysics." The ideas are not new, and are based on the combination of psychology, consumer research, and statistics (Moskowitz et al. 2006). The way of thinking offered by Mind Genomics allows one to explore ideas of commonplace issues, doing so rigorously, with results showing patterns that can be critical in application, patterns that are easy to discover (Kover et al. 2022). The theoretical foundation of Mind Genomics has been established in a number of publications, Moskowitz and Gofman (2007), Gofman and Moskowitz (2010), Gofman (2009, 2012), Moskowitz et al. (2001), and Porretta et al. (2019), to name a few.

Mind Genomics is used in many research fields to analyze complex issues in the field of education (Todri et al. 2020; Gere et al., 2019a; Meka and Papajorgji 2020; Moskowitz et al. 2023) in retail and supply chain field, (Ilollari et al. 2020a, b; Papajorgji et al. 2021) in the field of tourism, (Gere et al., 2018; Saulo et al. 2019) in the field of marketing (Moskowitz et al. 2020) in the field of law, (Gere et al. 2019c; Papajorgji and Moskowitz 2022) in the field of applied psychology (Gabay and Moskowitz 2015, 2020; Gabay et al. 2018) in the field of healthcare, etc.

The direct application of Mind Genomics thinking is already making itself known in the social and political science (Moskowitz et al. 2020; Papajorgji and Moskowitz 2022; Gere et al. 2019b).

1.4 The Contribution of This Book

What is the news in this book? The news in this book is that the marriage between Artificial Intelligence (AI) and quantitative research methods, specifically Mind Genomics, offers a powerful combination of data-driven insights and advanced analytics. This integration has transformed various fields, including finance, healthcare, marketing, and social sciences, by enhancing the accuracy, efficiency, and scope of quantitative research.

Case studies presented in this book highlight the practical application of AI in political science, showcasing real-world examples where AI has provided insights into political behavior, election outcomes, and policy analysis. Real-world examples of using combinations of AI and quantitative methods demonstrate how the book encourages innovative approaches to political science research, offering new and easy-to-implement methods for data-driven analysis and decision-making, without requiring a background in computer science. Topics include some of the challenges political scientists face, such as analyzing large datasets, predicting political trends, and understanding complex political dynamics.

While presenting the advantages of using AI and quantitative approaches in political sciences, the book addresses how to tackle ethical concerns related to AI, such as bias, privacy, and the impact of automation on political processes, guiding readers to use AI responsibly.

As large language models (LLMs) performance gets better and better every day, there are, though, some things that need to be clarified in terms of what they can and cannot do.

It is essential to state the function of LLMs one more time. They are word guessers; given a particular set of words as input, they will make the best guess for the next one. In terms of probability, LLMs find the word of higher probability given a set of input words (Blank 2023; Häggström 2023); they are pattern-matching. All uses of LLMs are designed and implemented around this simple idea. These models cannot consider the larger context in which a word appears and its multiple meanings and associations (Burtsev et al. 2023).

Large language models do not do problem-solving or planning (Riedl 2023). There are many examples in which people will ask LLMs to build a plan about something, and LLMs will respond in a way that may look like a perfect plan to many people. Plans created by LLMs may ignore particular details about the world and have the tendency to be the most generic plan. Large language models certainly have not "thought through the alternatives" or tried one idea and backtracked and tried another. There is no mechanism inside a transformer that one could point to that which would do such a back-and-forth consideration of the future (Valmeekam et al. 2022).

It needs to be strongly stated that LLMs do not have planning capabilities but have seen plenty of examples of plans during training phases, and they can match the words together to create sentences, which look like a plan. When humans make plans, they predict the outcome given a particular input. Humans have the ability to analyze, predict, and rationalize the results. LLMs do not have the ability to work toward a goal by making choices between alternatives that are likely to get closer to that goal. Another vital point to make is that LLMs do not have goals.

Thus, the literature suggests that every plan received as output from LLMs must be controlled carefully (Guerreiro et al. 2023; Azamfirei et al. 2023).

The swift advancement of artificial intelligence (AI) poses distinct difficulties for writers as it is a constantly evolving technology, with new applications and advancements appearing regularly. Information may quickly become outdated due to this speed, frequently even before it is published. Thus, the rapid change of AI presents a severe risk to authors, journalists, and academics since it could cause their analyses and insights to become less accurate or relevant, which would damage their credibility. Moreover, because AI technology is still developing, debates and projections on its future are made more difficult by the potential for significant shifts in societal impacts and ethical issues. As a result, talking about AI necessitates a current grasp of the subject and caution due to its explosive nature.

We must never forget the disclaimer of CHATGPT 4: "CHATGPT can make mistakes. Consider checking important information."

References

Azamfirei, R., Kudchadkar, S. R., & Fackler, J. (2023). Large language models and the perils of their hallucinations. *Critical Care*, *27*(1), 120. https://doi.org/10.1186/s13054-023-04393-x

Blank, I. A. (2023). What are large language models supposed to model? *Trends in Cognitive Sciences*, *27*(11), 987–989. https://doi.org/10.1016/j.tics.2023.08.006

Bruch, E., & Feinberg, F. (2017). Decision-Making Processes in Social Contexts. *Annual Review of Sociology*, *43*, 207. https://doi.org/10.1146/ANNUREV-SOC-060116-053622

Burtsev, M., Reeves, M., & Job, A. (2023). The Working Limitations of Large Language Models. *MAGAZINE WINTER 2024 ISSUE FRONTIERS*.

Gabay, G., & Moskowitz, H. R. (2015). Mind Genomics : What Professional Conduct Enhances the Emotional Wellbeing of Teens at the Hospital? *Journal of Psychological Abnormalities in Children Mind Genomics*, *January*. https://doi.org/10.4172/2329-9525.1000147

Gabay, G., Zemel, R., Gere, A., Zemel, G., Papajorgji, P., & Moskowitz, H. R. (2018). On the Threshold: What Concerns Healthy People about the Prospect of Cancer? *Cancer Studies and Therapeutics, Research Open*, *3*(4).

Gabay, Gillie, & Moskowitz, H. R. (2020). *"Are We There Yet ?" Mind-Genomics and Data-Driven Personalized Health Plans*. 7–28. https://doi.org/10.1108/978-1-83867-249-220191003

Gere, A., Zemel, R., Papajorgji, P., & Moskowitz, H. (2018). Weak Signals and Mind-Sets of Consumers: The Case of Milk. *Journal of Food Science and Engineering*, *8*, 125–136.

Gere, A., Papajorgji, P., Moskowitz, H. R., & Milutinovic, V. (2019a). Using a Rule Developing Experimentation Approach to Study Social Problems. *International Journal of Political Activism and Engagement*, *6*(3), 23–48. https://doi.org/10.4018/ijpae.2019070103

Gere, A., Shelley, R., Zemel, R., Papajorgji, P., Deitel, J., & Moskowitz, H. (2019b). The Mind Assesses Aggression – Russia vs the Ukraine: A Mind Genomics Exploration. *Ageing Science & Mental Health Studies*, *3*(1). https://doi.org/10.31038/ASMHS.2019315

Gere, A., Zemel, R., Papajorgji, P., & Moskowitz, H. (2019c). "Candy Is Dandy": The Mind of Sexuality as Suggested by a Mind Genomics Experiment. In B. Guthrie, J. D. Beauchamp, A. Buettner, S. Toth, & M. C. Qian (Eds.), *Sex, Smoke, and Spirits: The Role of Chemistry* (1st Editio, pp. 17–31). American Chemical Society. https://doi.org/10.1021/bk-2019-1321

Gofman, A. (2009). Extending Psychophysics Methods to Evaluating Potential Social Anxiety Factors. *Medicine*, *346*(17), 1337–1342.

Gofman, A. (2012). Origins of RDE and the Role of Experimentation in Consumer- Driven Innovation. In H. R. Gofman, A. Moskowitz (Ed.), *Rule Developing Experimentation: A Systematic Approach to Understand & Engineer the Consumer Mind* (pp. 3–33). Bentham ebooks. https://doi.org/10.2174/97816080528441120101

Gofman, A., & Moskowitz, H. (2010). ISOMORPHIC PERMUTED EXPERIMENTAL DESIGNS AND THEIR APPLICATION IN CONJOINT ANALYSIS. *Journal of Sensory Studies*, *25*(1). https://doi.org/10.1111/j.1745-459X.2009.00258.x

Guerreiro, N. M., Alves, D. M., Waldendorf, J., Haddow, B., Birch, A., Colombo, P., & Martins, A. F. T. (2023). Hallucinations in Large Multilingual Translation Models. *Transactions of the Association for Computational Linguistics*, *11*, 1500–1517. https://doi.org/10.1162/tacl_a_00615

Häggström, O. (2023). Are Large Language Models Intelligent? Are Humans? *IS4SI Summit 2023*, 68. https://doi.org/10.3390/cmsf2023008068

Heidelberger, M. (1994). The Unity of Nature and Mind: Gustav Theodor Fechner's Non-Reductive Materialism. In Romanticism in Science.

Ilollari, O., Papajorgji, P., & Civici, A. (2020a). Tourism in Albania and its challenges during and after Covid-19. *UNICART3, International Conference, Academic Research & Tourism.*

Ilollari, O., Papajorgji, P., & Civici, A. (2020b). Understanding client's feelings about mobile banking in Albania. *2 Nd UNICART – INTERDISCIPLINARY INTERNATIONAL CONFERENCE ON MANAGEMENT, TOURISM AND DEVELOPMENT OF TERRITORY*, 147–154.

Kover, A., Papajorgji, P., & Moskowitz, H. (2022). *Applying Mind Genomics to Social Sciences*. IGI Global. https://doi.org/10.4018/978-1-7998-8409-5

Meka, E., & Papajorgji, P. (2020). Financial Literacy on EU Financial Infrastructure and Institutions: Points of View from the Albanian Students, using the Mind Genomics Approach. *3rd UNICART INTERDISCIPLINARY INTERNATIONAL CONFERENCE*, 229–238.

Moskowitz, H. ., Todri, A., Papajorgji, P., Rappaport, S., & Deitel, Y. (2023). Sourcing and Vetting Ideas for Sustainability in the Retail Supply Chain: The Contribution of Artificial Intelligence Coupled with Mind Genomics. *NTERNATIONAL JOURNAL ON FOOD SYSTEM DYNAMICS*, *14*(4), 367–380. https://doi.org/ps://doi.org/10.18461/ijfsd.v14i4.H1

Moskowitz, H., Beckley, J., & Ashman, H. (2006). Founding A New Science: Mind Genomics. *Journal of Sensory Studies*. https://doi.org/10.1111/j.1745-459X.2004.00066.x

Moskowitz, H., & Gofman, A. (2007). *Selling blue elephants: How to make great products that people want before they even know they want them.* Pearson Education.

Moskowitz, H., Gofman, A., Itty, B., Katz, R., Manchaiah, M., & Ma, Z. (2001). Rapid, inexpensive, actionable concept generation and optimization: the use and promise of self-authoring conjoint analysis for the food service industry. *Blackwell Science Ltd. 2001 Food Service Technology, 1, pp. 149–167*, *1*, 19.

Moskowitz, H., Wren, J., & Papajorgji, P. (2020). *Mind Genomics and the Law* (1st Editio). LAP LAMBERT Academic Publishing.

Ortlieb, S. A., Kügel, W. A., & Carbon, C.-C. (2020). Fechner (1866): The Aesthetic Association Principle—A Commented Translation. I-Perception, 11(3), 204166952092030. https://doi.org/10.1177/2041669520920309

Papajorgji, P., Ilollari, O., Civici, A., & Moskowitz, H. (2021). A Mind Genomics-Based Cartography to Assess the Effects of the COVID19 pandemic in the tourism industry. *WSEAS TRANSACTIONS ON ENVIRONMENT AND DEVELOPMENT*, *17*, 1021–1029. https://doi.org/10.37394/232015.2021.17.94

Papajorgji, P., & Moskowitz, H. (2022). The 'Average Person' Thinking About Radicalization: A Mind Genomics Cartography. *Journal of Police and Criminal Psychology*. https://doi.org/10.1007/s11896-022-09518-6

Porretta, S., Gere, A., Radványi, D., & Moskowitz, H. (2019). Mind Genomics (Conjoint Analysis): The new concept research in the analysis of consumer behaviour and choice. *Trends in Food Science and Technology*, *84*(January), 29–33. https://doi.org/10.1016/j.tifs.2018.01.004

Riedl, M. (2023). *A Very Gentle Introduction to Large Language Models without the Hype*.

Saulo, A., Moskowitz, V., Gere, A., Papajorgji, P., & Ettinger Lieberman, L. Feuerwerker, D. (2019). Linking Food Endorsement Labels & Messaging to Perceived Price and Emotions. A Mind Genomics ® Exploration. *Advances in Nutrition and Food Science*, *2019*(5).

Stevens, S. S. (1975). Psychophysics: Introduction to its perceptual, neural, and social prospects. John Wiley & Sons Inc.

Todri, A., Papajorgji, P., Moskowitz, H. R., & Scalera, F. (2020). Perceptions regarding Distance Learning in Higher Education, Smoothing the Transition. *Contemporary Educational Technology*, *12*(1). https://doi.org/10.30935/cedtech/9274

Valmeekam, K., Marquez, M., Olmo, A., Sreedharan, S., & Kambhampati, S. (2022). *PlanBench: An Extensible Benchmark for Evaluating Large Language Models on Planning and Reasoning about Change*. http://arxiv.org/abs/2206.10498

Chapter 2
Analytic Essentials: Linear Regression

Many books explain linear regression quite well, appropriate for the different levels of reader comprehension, e.g. Weiss (2012), Nikoletseas (2020), and Frost (2020) to name a few. So important is linear regression to this book that we have taken the liberty of devoting a chapter to it to ensure that all readers begin with at least a rudimentary understanding.

Chapter 3, 4, 5, 6, 7, 8, 9, and 10, the remainder of the book, will introduce statistical concepts, which build upon linear regression. Our desire is simply to give a foundation to this important idea. We never lose sight of the fact that our readers include social science and political science students, researchers, and anyone who wants to use quantitative methods in their daily work, each with a different type of appreciation of linear regression, and its world of capabilities.

2.1 Introduction to the Concept of Linear Regression

It is common knowledge that any manager or policymaker should make data-driven judgments whenever feasible. How is one able to sort through all of the facts available? Nowadays, people generally do not have to perform the math calculations themselves. However, it is still necessary to comprehend and evaluate the study and the results computer systems have produced accurately. Regression analysis is one of the most significant categories of data analysis.

Every sales manager must deal with analysis, including forecasting the figures for the following month. The management is aware that the results can be impacted by dozens, if not hundreds, of factors, such as the weather, a competitor's promotion, or rumors of a new and superior model. Someone else in the company may have an idea about what will impact sales the greatest. One possibility could be that more merchandise is sold when it rains, and much more merchandise is sold when

P. Papajorgji, H. Moskowitz, *The Mind of Everyday*,
https://doi.org/10.1007/978-3-031-78078-3_2

it rains a lot. Another viewpoint would be that sales increase about a month following the competitor's offer.

Regression analysis is the process of quantitatively determining which of those aspects (also called variables) has an effect. What matters most? Can we overlook any of the factors? What interactions exist between those factors?

In statistics, those factors are called "variables." Usually, there is one *dependent variable* in a study—the main factor that we are trying to understand or predict. In the sales example above, the dependent variable is monthly sales. Also, there *are independent variables*—the factors suspected to impact the dependent variable.

To conduct a regression analysis, one must gather data regarding variables in question. The manager does not have to gather the data but needs to understand the data collection process. Thus, the manager may ask for the monthly sales numbers for the past 4 years and any data on the independent variables, which may help shed light on the topic. So, in this case, the manager will find out the average monthly rainfall for the past 4 years to learn whether or not rainfall helps to increase sales.

The sales and rainfall data are shown in Fig. 2.1. Is there a connection between these data, the management would wonder? Do these data have any relation with each other? Does the daily rain have an impact on sales?

The dependent variable, or the item the manager is interested in, is always on the Y-axis, which is the quantity of sales, and the X-axis is the total amount of rainfall. Each filled circle shows data: The amount of rain and the number of sales in a given month.

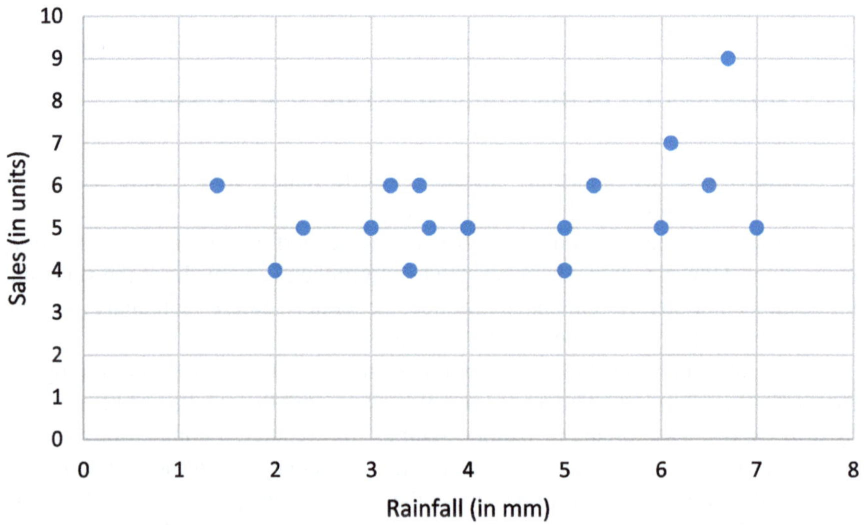

Fig. 2.1 Sales and rainfall data

2.1.1 Linear Equations

Looking at these data, days with much rain had higher sales. Knowing that is interesting, but to what extent? If it rains nine millimeters, how much will the sales be? How about 10 mm of rain? Imagine drawing a line through the middle of the data points in the preceding chart. This line will help us determine the average sales amount after a specific amount of rain, as shown in Fig. 2.2.

This line that we just drew is called the "regression line." The regression line is typically drawn automatically with today's programs (e.g., SPSS or even Excel), with the line closest to the actual points. In other words, the red line is the best explanation of the relation between the independent variable and the dependent variable.

Before the days of computers, however, it was a difficult task to create the line using statistical methods, and the line was often fit by eye (the method of "ocular trauma" in the words of Harvard University graduate students in the 1960s.

In addition to drawing the line, the statistics program also outputs a formula, which explains the slope of the line, expressed as:

$$Y = b0 + b1X$$

where $b0$ and $b1$ are constants (fixed numbers), X is the *independent* variable, and Y is the *dependent* variable.

A linear equation with a single independent variable has a graph, which is a straight line, or just a line. Applications of mathematics to a wide range of

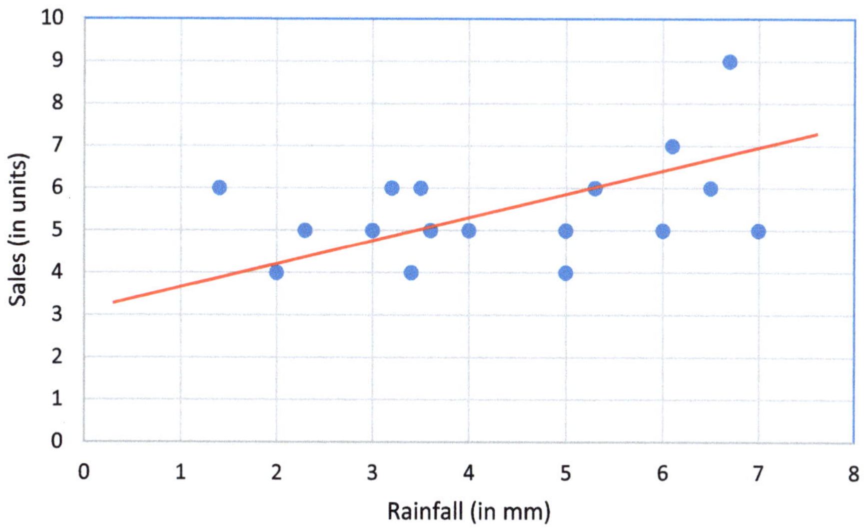

Fig. 2.2 A line passes through the data points

disciplines, including the physical and mathematical sciences, management, life, and social sciences, usually use linear equations with a single independent variable.

Consider this example. We have a factory that produces a product. The production cost comprises two elements: the worker's time on the job and a fixed cost covering different production elements. For example, assume that when an employee works for 1 h, the employee will be paid 15$; if 2 h, then will be paid 30$. The amount of money received depends on the amount of time spent working. Assuming that the fixed costs are 45$, the total production cost will equal the fixed cost + the worker's time on the job. Thus, using a linear equation, the cost of the production can be expressed as

$$Y = 45 + 15x$$

where $b0 = 45$ and $b1 = 15$.

When the person works 2 h the cost will 75$. When the person works 3 h, the cost will be 90$. Table 2.1 shows the numbers for a full day of work.

Figure 2.3 shows the graphic presentation of values in Table 2.1. This graph can be used to get a quick idea of the cost. For example, working for 3.5 h generates a total cost of around 100, $97.5 to be exact. Figure 2.3 shows that when time spent in production increases, so does the production cost. Table 2.1 shows that when the working time increases by an hour, the production cost increases by $15.

2.1.2 *Intercept and Slope*

The value of the line at the place where the Y-axis and the line cross is represented by the number $b0$ in the linear equation $Y = b0 + b1X$. The number $b1$ indicates the steepness of the line. *In more practical terms, more accurately, b1 shows how much the Y-value changes when the X-value increases by one unit.*

Figure 2.4 graphically shows that when the independent variable increases by one unit, the value of $Y = b0 + b1X$ increases by $b1$ units. In statistics, the parameter $b0$ is called the Y-intercept (or just the intercept), and $b1$ is called the slope.

Figures 2.3 and 2.5 have different directions, as can be seen if we compare them. Since the slope of the line in Fig. 2.3 is positive ($b1 > 0$), it is evident that the line

Table 2.1 Data regarding the hours and the corresponding cost

Hours	Cost ($)
1	60
2	**75**
3	**90**
4	**105**
5	**120**
6	**135**
7	**150**

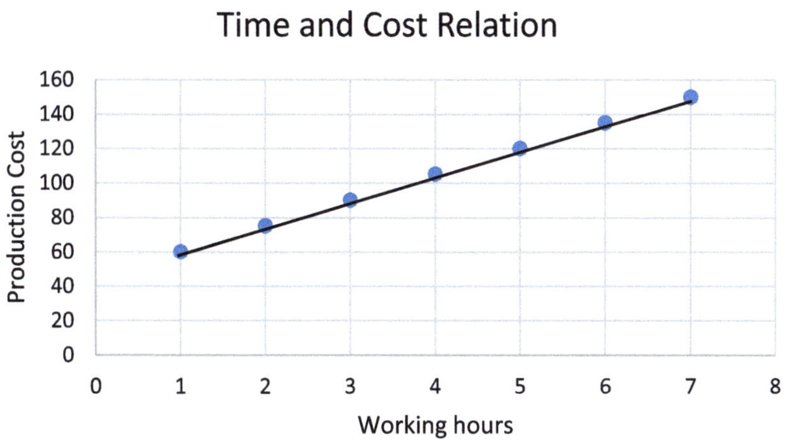

Fig. 2.3 The straight line shows the relation between hours and cost of production

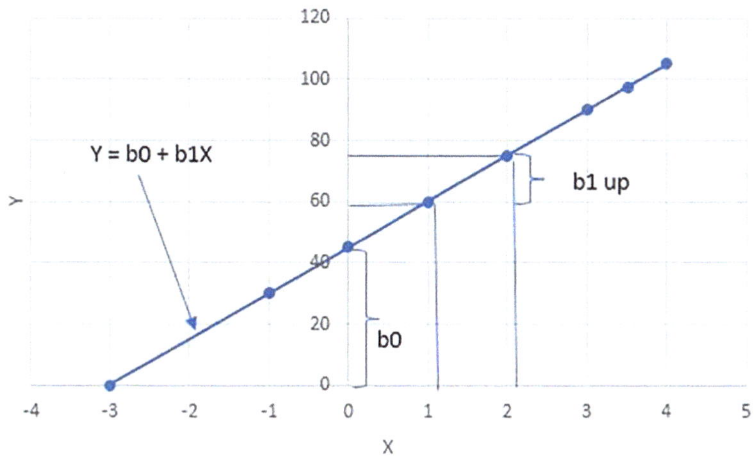

Fig. 2.4 Slope and intercept

slopes upward and the Y-values rise as X increases. Since $b1 = -3 < 0$, the line in Fig. 2.5 has a negative slope, meaning that as X increases, the Y-values drop.

Note: The graph of the linear equation $Y = b0 + b1X$ slopes upward when $b1 > 0$, slopes downward if $b1 < 0$, and is horizontal, parallel to the X-axis when $b1 = 0$

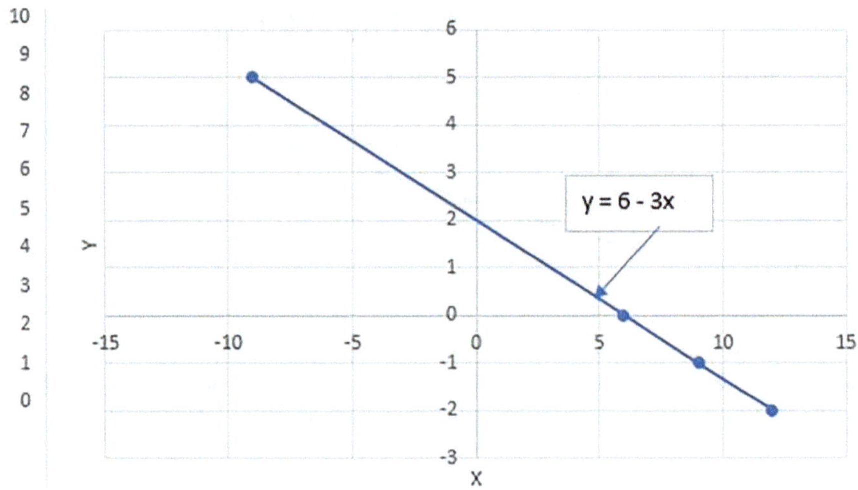

Fig. 2.5 Line with negative slope

2.2 The Regression Equation

Seldom are real-world applications as straightforward as the production-cost example, wherein the value of one variable can be accurately predicted using the value of another variable. Instead, we frequently depend on imprecise forecasts. For example, knowing an apartment's surface area in square meters (X) does not allow us to determine its actual asking price (Y). Prices differ among apartments, even for apartments with the same size, e.g., 90 m^2. Unless we have other information to help us, when we have to estimate the cost of a 90-m^2 apartment we will have to settle for an *estimate* of the average or mean price, a so-called "ballpark estimate."

Let us look at the homework of a buyer, who assembles data about apartments before making an offer. Our buyer has access to information about the size of the apartment and the price of the apartment. Our buyer's collected data on the apartment's surface and the corresponding prices appear in Table 2.2.

To get a sense about the relation between apartment size and price, we plot the data from Table 2.2 in a graph as shown in Fig. 2.6. The points are scattered irregularly; no line joins all the points. There is variability, scatter. Yet we do see a pattern, although not a perfect pattern. The points, the prices, seem to fall along a line, albeit with scatter. Our buyer will fit a line to the data points and use that line to predict the price of an apartment based on its surface.

We want a way to select the "best" line from the numerous lines that might be drawn through the data point cluster. The process, known as the *least-squares criterion*, is predicated on creating a line with the property that we minimize "error," viz., get the line as "close" to the points as possible.

Our strategy is simple. We will adopt a specific criterion and adjust the line until we satisfy the criterion. The criterion is to define "departure from the line" as the

Table 2.2 Surface (in square meters) and prices (in 100's of dollars)

Apartments	Surface	Price
1	60	850
2	65	950
3	70	940
4	75	1000
5	80	1010
6	85	1130
7	90	1150
8	90	1220
9	105	1350
10	115	1400
11	**130**	**1600**

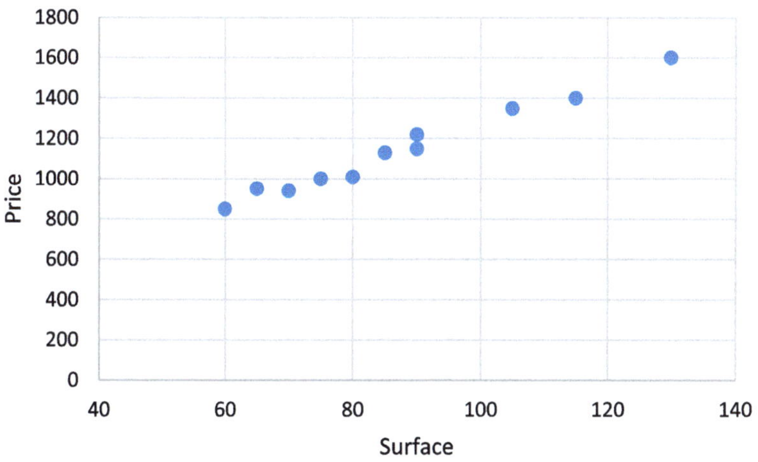

Fig. 2.6 Distribution of points (surface, price)

square of the difference between the actual value of the point and the estimated value of the point that the line would predict. By moving the line around we will create different values of this "*square of the difference between the actual value of the point and the estimated value of the point.*" Now to find the one line which MINIMIZES this error or sum of squares of departures from the line.

Across the many lines, which we can create to go through the dots in Fig. 2.7, we have selected only two to demonstrate the least-squares criterion. To find the equations for the two lines that will be used to demonstrate the principle of the least-squared criterion, we will write the system of two linear equations for the values of X and Y from Table 2.2 and solve the system. We will consider the point ($X = 65$, $Y = 950$) and the point ($X = 90$, $Y = 1220$). The system is:

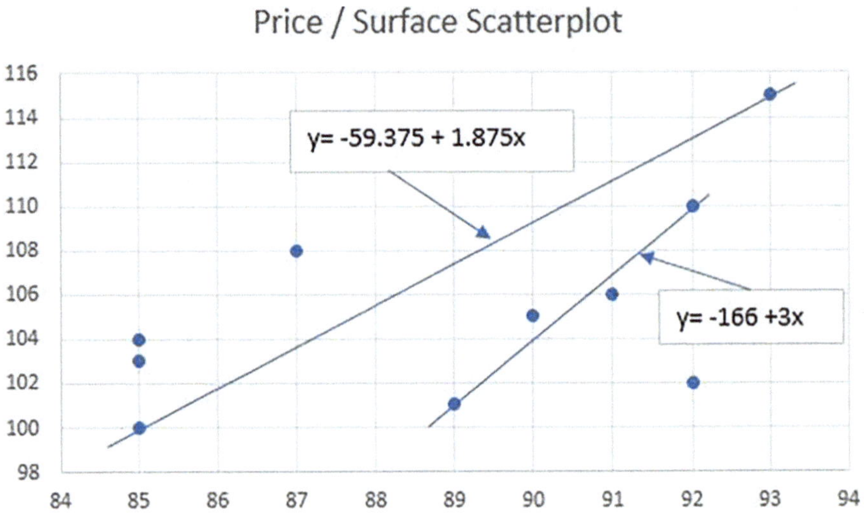

Fig. 2.7 Lines which could "fit" the dots of the graph

$$\begin{cases} 950 = b0 + 65b1 \\ 1220 = b0 + 90b1 \end{cases}$$

We provided CHATGPT 4 (OpenAI 2024) with these data and received the solution immediately: $b0$ = 248 and $b1$ = 270/25. Thus, the equation of the first line (L1) is:

$$(\text{L1}) \quad Y = 248 + 270/25\, X$$

and for the second (L2) we considered two points (X = 75, Y = 1000) and (X = 115, Y = 1400). Thus, the system of two equations is:

$$\begin{cases} 1000 = b0 + 75b1 \\ 1400 = b0 + 115b1 \end{cases}$$

To solve this system is very simple, we subtract the second equation from the first one and calculate $b1$. Next, use the value of $b1$ in any of the equations to calculate $b0$. So, we have $b0$ = 250 and $b1$ = 10. Thus, the second line, equation is:

$$(\text{L2}) \quad Y = 250 + 10\, X.$$

Let us consider first L1. $\hat{y}$ is the value calculated using equation L1, and Y is the value given by Table 2.2 for a given value of X. Thus, the difference $e = Y - \hat{y}$ provides the error caused by using $\hat{y}$ instead of Y.

Table 2.3 Preparing elements for calculating the "best" fit between L1 and L2

Line Y = 248 + 270/25 X						**Line Y = 250 + 10 X**				
X	Y	ŷ	e	e²		X	Y	ŷ	e	e²
60	850	896	-46	2116		60	850	850	0	0
65	950	950	0	0		65	950	900	50	2500
70	940	1004	-64	4096		70	940	950	-10	100
75	1000	1058	-58	3364		75	1000	1000	0	0
80	1010	1112	-102	10404		80	1010	1050	-40	1600
85	1130	1166	-36	1296		85	1130	1100	30	900
90	1150	1220	-70	4900		90	1150	1150	0	0
90	1220	1220	0	0		90	1220	1150	70	4900
105	1350	1382	-32	1024		105	1350	1300	50	2500
115	1400	1490	-90	8100		115	1400	1400	0	0
130	**1600**	1652	-52	2704		**130**	**1600**	1550	50	2500
Sum of errors: 38004						**Sum of errors: 15000**				

If we consider the point ($X = 90$, $Y = 1150$), values from Table 2.2, and the value Y obtained by replacing X in the equation (L1) $Y = 248 + 270/25\ X$ will have the value $\hat{\mathbf{y}} = 1220$ obtained from the prediction line. Thus, the error e at this point is $e = Y - \hat{\mathbf{y}} = -70$ and $e^2 = 4900$ as shown in Table 2.3 in the highlighted cells. Using Excel, it is easy to obtain values for the entire table in no time.

Between many lines, which may go through the dots in Fig. 2.7, we have selected only two to demonstrate the least-squares criterion.

The first line (L1) is $Y = -59.375 + 1.875\,X$, and the second (L2) is $Y = -166 + 3\,X$.

Let us consider first L1. $\hat{y}$ is the value calculated using equation L1, and Y is the value given by Table 2.2 for a given value of X. Thus, the difference $Y - \hat{y}$ provides the error caused by using $\hat{y}$ instead of Y. Table 2.3 presents all the elements, which make calculations easy using Excel.

Table 2.3 presents all the elements, which make calculations easy using Excel. In Table 2.2, we compute the sum of the squared errors for both lines presented in columns e^2, to determine which line, Line 1 or Line 2, better fits the data. The line that best matches the data is the one with the smallest total of squared errors, in this case, Line 1. The least-squares criterion states that the line that best fits the data is the one with the fewest sum of squared errors among all other lines.

Note: The least-squares criterion is that the line that best fits a set of data points has the smallest possible sum of squared errors.

As shown in Table 2.3 the smallest sum of errors (15,000 versus 38,004) is provided by line **Y = 250 + 10 X**, thus we conclude that this line is a better fit between the two lines.

Table 2.4 Formulas for regression and correlation

Quantity	Defining formula	Computing formula
S_{xx}	$\Sigma(x_i - \bar{x})^2$	$\Sigma x_i^2 - (\Sigma x_i)^2 / n$
S_{xy}	$\Sigma(x_i - \bar{x})(y_i - \bar{y})$	$\Sigma x_i y_i - (\Sigma x_i)\ (\Sigma\ y_i/n$
S_{yy}	$\Sigma(y_i - \bar{y})^2$	$\Sigma y_i^2 - (\Sigma y_i)^2 / n$

Source Weiss (2012)

Table 2.5 Facilitating the calculations of *b*0 and *b*1

X	*Y*	*XY*	X^2	$\bar{X}$	$\bar{Y}$
60	850	51,000	3600	87.72727	1145.455
65	950	61,750	4225		
70	940	65,800	4900		
75	1000	75,000	5625		
80	1010	80,800	6400		
85	1130	96,050	7225		
90	1150	103,500	8100		
90	1220	109,800	8100		
105	1350	141,750	11,025		
115	1400	161,000	13,225		
130	**1600**	208,000	16,900		
965	**12,600**	**1,154,450**	**89,325**		

The regression line for a given collection of data points must satisfy a specific condition, but the least-squares criterion does not specify how to find that line. Finding the line is achieved via formula in Table 2.4. We introduce the basic terminology used in our regression and correlation analyses.

The regression equation for a set of *n* data points is $\hat{y} = b0 + {}_{b1X}$, where

$$b1 = \frac{S_{xy}}{S_{xx}} \text{ and } b0 = \frac{1}{n}\left(\Sigma \mathbf{y_i} - b1 \Sigma \mathbf{x_i}\right) = \bar{y} - b1\bar{x}$$

Now that we have the formulas shown in Table 2.4, we can draw the regression line. First, we need to compute b1 and b0 using the formulas presented in Table 2.4 and the values for *X*, *Y*, *xy*, x^2, and their sums presented in Table 2.5.

Thus,

$$b1 = \frac{S_{xy}}{S_{xx}} = 10.51509$$

and

$$b0 = \bar{y} - b1^* \bar{x} = 1145.455 - 10.51.509^* 87.72727 = 222.9944$$

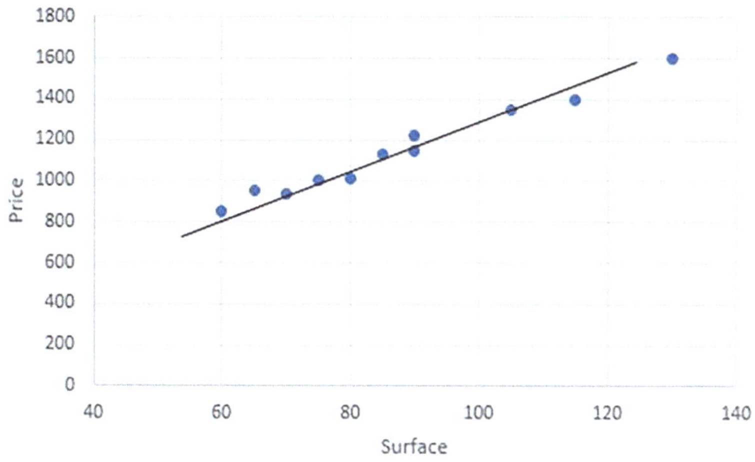

Fig. 2.8 The regression line

Thus, the regression line is **$\hat{y}$ = 222.9944 + 10.51509 X**.

Note: Rounding cautions are applicable as normal. Do not round the results until the computation is complete when determining the regression line's slope, $b1$. It is advised not to utilize the rounded value of $b1$ when calculating the Y-intercept, b0; instead, maintain full calculator accuracy.

We must choose two points and connect them with a line in order to draw the regression line. Consequently, P1(X = 70, Y = 1004) and P2(X = 105, Y = 1382) are the two points. The regression line is depicted in the Fig. 2.8, crossing points P1 and P2. Figure 2.8 also displays the other points (X,Y) in Table 2.3.

The same result is reached using Excel's ***Analyze Data*** tab. The X and Y columns must be selected by choosing the ***Regression*** option in the list of available options. The regression line and the value of R^2 are shown in Fig. 2.9.

It should come as no surprise that price tends to climb as surface increases given the regression line's positive slope ($b1$). Given that Y is the price in hundreds of dollars and X is the surface area in square meters, the slope of 10.51509 suggests that the price increases by an estimated \$1051 per square meter, at least for areas between 60 and 130 square meters. A 90-m^2 apartment would roughly cost \$116,900.

2.2.1 Some Terminology

The ***dependent*** variable in the linear equation $Y = b0 + b1x$ is X, and the ***independent*** variable is Y. Typically, X is used to predict or explain the values of the response variable, so in the context of regression analysis we refer to it as the independent variable or the predictor. For our example, the independent variable is surface area, the dependent variable is price.

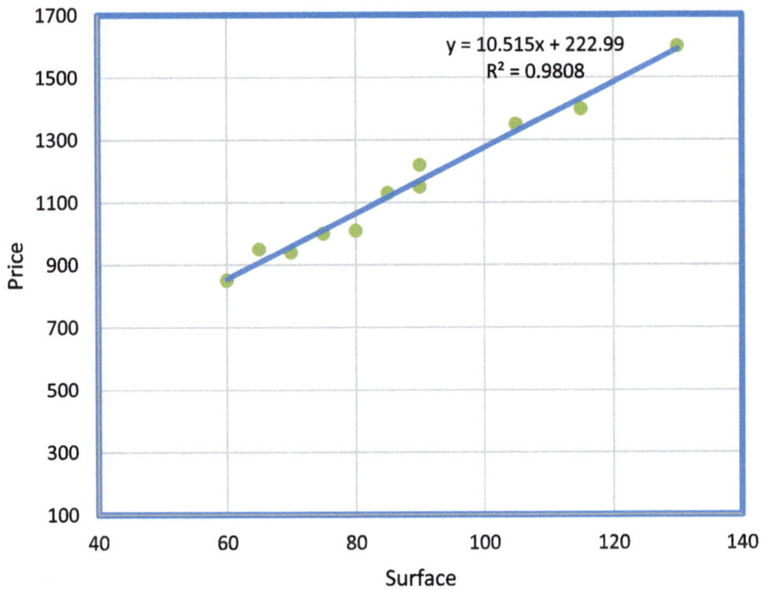

Fig. 2.9 The regression line, the equation, and the value of r^2

Here is a worked example to illustrate the point. First, let us assume that two variables have a linear relation, which we see in the scatterplot (Fig. 2.10). In this simple case the regression equation adequately represents the price within the observed values of the predictor variable. As a result, we can make predictions of price when we know space (viz., when we know the value for the independent variable we can accurately predict the value of the dependent variable. ***Extrapolating*** outside of that range, as in this case, may or may not be accurate, although it's more likely to be accurate when the data points are close to the regression line. When we move beyond the range of the actual data to a new region, higher or lower than the data we used, we really have no idea of the relation.

In our case, taking an $X = 150$ outside the initial range (60,130) will give a value of $180,025 for the apartment price, which looks reasonable price. In this case extrapolation provides a reasonable estimate. Let's turn the problem around. What about a small surface area of 40? Is it likely to be around 600? Should we be confident?

An ***outlier*** is an observation that deviates from the general pattern. An outlier in the context of regression is a data point, which, in comparison to the other data points, is located far from the regression line. Regression analysis can occasionally be significantly affected by an outlier. As such, it is necessary to detect outliers and eliminate them from the analysis when appropriate, such as when an outlier is determined to be a measurement or recording error. There are no outliers in our example in Figs. 2.8 and 2.9.

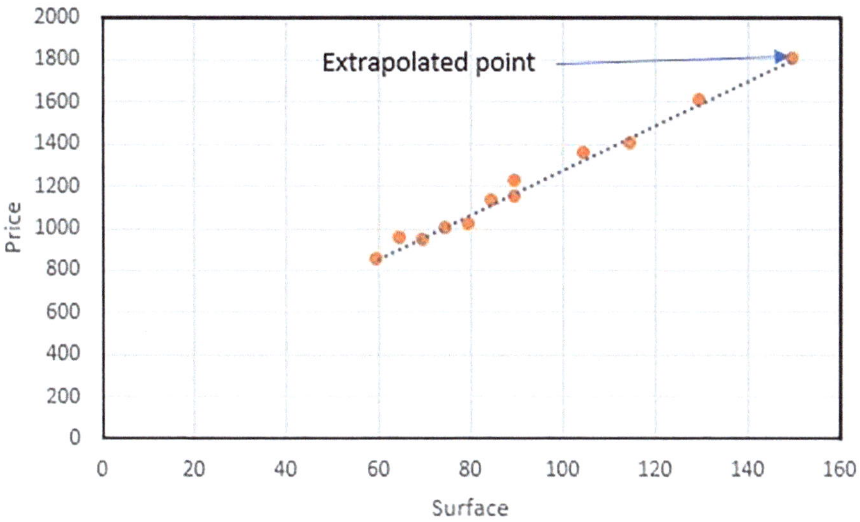

Fig. 2.10 Adding an extrapolated point

A crucial first step is to keep an eye out for a data point, which significantly alters the regression equation (and line) after it is removed from the regression analysis. This point is considered to be ***influential*** in regression analysis. The word "influential" is used because the regression line is "pulled" toward a data point, which is separated from the other data points in the X-direction without any opposition from the other data points; this type of data point is frequently seen as important.

An influential observation can be eliminated though if it is the result of a measurement or recording error, or if it is evident for some other reason that it should not be in the dataset. Nonetheless, it can be challenging to decide whether to keep an influential observation when there is no explanation for it, necessitating the researcher's judgment.

Influential observations can be outliers or not, and outliers can also be influential observations. Many statistical software applications provide the means to identify influential observations and possible outliers.

Assuming that the data points are dispersed over a line is the foundation for the concept of identifying a regression line. Often, though, the data points are not distributed along a line but rather along a curve. For these data points, it is still possible to construct $b0$ and $b1$ to get a regression line. However, the outcome will result in an improper line fit when a curve should have been utilized instead. For instance, the regression line suggests that the Y-values will keep increasing even though they have begun to decrease.

Note: **It is good practice to graph the data points as a scatterplot before attempting to construct a regression line for them. It may be misleading to estimate a regression line when the data points do not seem to be distributed about a line.**

2.3 The Coefficient of Determination

In the apartment example, we found the regression equation $\hat{y} = 222.9944 + 10.51509*X$ for the price and apartment surface data of a sample of 11 apartments, where $\hat{y}$ denotes the expected price in hundreds of dollars and X stands for the apartment surface in square meters. The price of an apartment measuring 110 m^2 was likewise predicted using the regression equation: = 222.9944 + 10.51509 * 110 = 137,965, or $137,965. However, how useful are these forecasts? Is it possible to estimate pricing using the regression equation, or would disregarding surface work just as well?

(Note: Often regression equations are written with many decimal points. The reality is that most people stop looking at the parameters of the equation beyond the first significant decimal digit after the decimal point).

There are several ways to assess a regression equation's suitability to make predictions. One technique calculates the percent of variation in the observed values, which is explained by the regression (or predictor variable). In our case this means the percent of variation in the prices of the apartments that we know that can be explained by the size of the apartment. To determine this percentage, we create two measures of variation:

1. The total variance in the response variable's observed values, and
2. The amount of variation in the response variable's observed values that the regression explains.

2.3.1 *Sums of Squares and Coefficient of Determination*

The sum of squared deviations between the observed values of the response variable and the mean or average of the response variable is used to calculate the overall variation in those values. The total sum of squares, or SST, is the name given to this measure of variance. SST = $\sum(y_i - \bar{y})^2$. The sample variance of the observed values of the response variable is obtained by dividing SST by $n - 1$.

First, we examine a specific observed value of the response variable, say, corresponding to the data point (x_i, y_i), as illustrated in Fig. 2.10, in order to calculate the amount of variation in the observed values of the response variable, which is explained by the regression. The divergence of each observed value from the mean value, $\sum(y_i - \bar{y})$, determines the overall variation in the observed values of the response variable. Each of these deviations, as illustrated in Fig. 2.10, may be broken down into two components: the deviation, which the regression line explains, $\hat{y}_i - \bar{y}$, and the residual deviation, which is not explained, $y_i - \hat{y}_i$. Therefore, the amount of variation (squared deviation) in the response variable's observed values, which the regression can account for is $\sum(\hat{y}_i - \bar{y})^2$. The regression sum of squares, or SSR, is the name given to this variation measure. SSR = $\sum(\hat{y}_i - \bar{y})^2$ as a result. We may calculate the percentage of variation in the observed values of the response

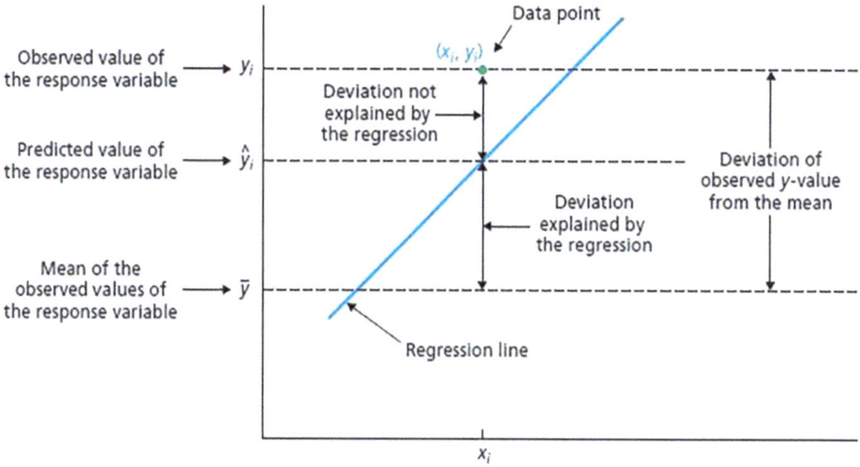

Fig. 2.11 Decomposing the deviation of an observed *Y*-value. (Source Weiss (2012)

variable—SSR/SST—which is explained by the regression using the total sum of squares and the regression sum of squares. This number, which goes by the symbol $\mathbf{r}^2$, is known as the coefficient of determination. Thus, SSR/SST = $\mathbf{r}^2$.

Let's look at the residual deviation shown in Fig. 2.11 before adding the coefficient of determination: the deviation that $y_i - \hat{y}_i$ cannot be explained by the regression. $\sum(y_i - \hat{y}_i)^2$ is the amount of variation (squared deviation) in the response variable's observed values, which the regression is unable to explain. The error sum of squares, or SSE, is the name given to this variation metric. Consequently, SSE = $\sum(y_i - \hat{y}_i)^2$.

Note: The range of values for the coefficient of determination, r^2 is 0 to 1. A regression equation's usefulness for making predictions is shown by a value of r^2 close to 1, whereas a value near 0 indicates that the regression equation is not very useful at all.

Thus, we can present some useful definitions as follows.

Total sum of squares, or ***SST***: The overall fluctuation in the response variable's observed values SST is equal to $\sum(y_i - \bar{y})^2$.

Regression sum of squares, or ***SSR***: The percentage change in the response variable's observed values, which the regression explains is:

$$\text{SSR} = \Sigma\left(\hat{y}_i - \bar{y}\right)^2.$$

Error sum of squares, or ***SSE***: The variance in the response variable's observed values, which the regression is unable to explain SSE, is equal to $\sum(y_i - \hat{y}_i)^2$.

Note that Table 2.6's fourth column's items were calculated with $\bar{y}$ = 1145.455. The fifth column's elements are computed using the same $\bar{y}$, and the regression sum of squares is obtained: The variation in the observed prices explained by the regression is SSR = $\sum(\hat{y}_i - \bar{y})^2$ = 516,147.4086.

Table 2.6 Helping calculate r^2

X	Y	$(Y-\overline{y})^2$	$\hat{y}$	$(\hat{y}-\overline{y})^2$
60	850	87,293.39	853.8998	85,004.1696
65	950	38,202.48	906.47525	57,111.10366
70	940	42,211.57	959.0507	34,746.3936
75	1000	21,157.02	1011.62615	17,910.03943
80	1010	18,347.93	1064.2016	6602.041145
85	1130	238.843	1116.77705	822.3987455
90	1150	20.66116	1169.3525	571.1122315
90	1220	5557.025	1169.3525	571.1122315
105	1350	41,838.84	1327.07885	32,987.388
115	1400	64,793.39	1432.22975	82,240.01794
130	**1600**	**206,611.6**	1589.9561	**197,581.632**
SST = 526,272.7			**SSR = 516,147.4086**	

We calculate the coefficient of determination, or the percentage of variation in the observed prices explained by the regression (i.e., by the surface-price linear relation for the sample of apartments), from SST and SSR:

$$r^2 = \frac{\text{SSR}}{\text{SST}} = 516147.4086 / 526272.7 = 0.98076 \text{ or rounded } 0.9808$$

which is the same value obtained by Excel-based calculations.

Given that 98% of the variation in the observed prices can be explained by the price regression on the apartment surface, it appears that the surface is indeed a useful tool for price prediction.

Also, it is useful to compute the error sum of squares for the data. We require both the observed and expected prices in order to compute SSE. Table 2.7 shows the values needed to calculate SSE.

The variation in the observed prices not explained by the regression is SSE = $\sum(yi - \hat{y}_i)^2$ = 10,125.07. SSE is the minimal sum of squared errors, which may be obtained among all lines since the regression line is the one which, in accordance with the least squares criterion, best fits the data.

SST = 526,272.7, SSR = 516,147.4086, and SSE = 10,125.07 pertain to the apartment data. We can see that SST = SSR + SSE since 526,272.7 = 516,147.4086 + 10,125.07 (taking the rounding impact into account). This is the regression identity equation, and it is always true.

One component of the overall variance in the observed values of the response variable represents the variation explained by the regression. In contrast, the other component represents the variation, which is not explained by the regression.

Note: SST = SSR + SSE

Thus,

Table 2.7 Helping calculate SSE

X	Y	$\hat{y}$	$Y - \hat{y}$	$(Y - \hat{y})^2$
60	850	853.8998	−3.8998	15.20844
65	950	906.4753	43.52475	1894.404
70	940	959.0507	−19.0507	362.9292
75	1000	1011.626	−11.6262	135.1674
80	1010	1064.202	−54.2016	2937.813
85	1130	1116.777	13.22295	174.8464
90	1150	1169.353	−19.3525	374.5193
90	1220	1169.353	50.6475	2565.169
105	1350	1327.079	22.92115	525.3791
115	1400	1432.23	−32.2298	1038.757
130	**1600**	1589.9561	10.0439	100.8799
SSE = 10,125.07				

$$r^2 = \frac{\text{SSR}}{\text{SST}} = \frac{\text{SST} - \text{SSE}}{\text{SST}} = 1 - \frac{\text{SSE}}{\text{SST}}$$

This calculation demonstrates that the coefficient of determination, when represented as a percentage, can also be understood as the percentage decrease in the total squared error, which results from utilizing the regression equation to predict the observed values of the response variable rather than the mean, $\bar{\mathbf{y}}$.

2.4 Basic Ideas About Correlation(s)

A statistical technique called correlation is employed to evaluate the likelihood of a linear relation between two continuous variables. It is easy to compute and understand. Nevertheless, correlation is so frequently abused by researchers that several statisticians have expressed their wish that the technique had never been developed in the first place (Mukaka 2012).

The biological sciences extensively use correlation coefficients. Although they can be used only for basic exploratory analysis or to build association networks for visualization, they are also essential to more complex multivariate data analysis techniques. As a result, having accurate correlation coefficient estimates is crucial. Comprehensive measurement techniques are employed in the current life sciences to quantify gene expressions, proteins, metabolites, and other forms of data (Saccenti et al. 2020).

Pearson and Spearman are the two correlation indices, which are most commonly employed. The correct usage of correlation coefficient type depends on the types of variables being studied (Mukaka 2012). This author also states that extreme values impact the Pearson coefficient, making it susceptible to exaggerating or weakening the strength of the association. When one or both of the variables are not normally distributed, it is, therefore, improper.

The first is used when the data has a normal distribution and calculates the linear relation between two continuous random variables. It is crucial to note that normality is essential for the calculation of the significance and confidence intervals, not the correlation coefficient itself (Akoglu 2018).

Conversely, the second approach, the Spearman approach, is utilized when the data do not conform to a normal distribution and quantifies any monotonic relation between two continuous random variables (Rovetta 2020).

2.5 The Pearson Product-Moment Correlation

During data analysis, we often hear analysts say that these two things correlate or do not correlate with each other. The correlation between two quantitative variables can be calculated using a variety of statistics. The most often used statistic is the linear correlation coefficient, or **r**, which is also known as the Pearson product-moment correlation coefficient, after Karl Pearson, who developed it.

The linear correlation must be calculated to conclude whether the two data series correlate. Thus, in data analysis, linear correlation measures the strength and direction of a linear relation between two variables. It helps in understanding how one variable changes in relation to another. The **r** definition formula is as follows:

$$\boldsymbol{r} = \frac{\frac{1}{n-1}\Sigma\left(xi-\overline{x}\right)\left(yi-\overline{y}\right)}{SxSy}$$

where Sx and Sy are the standard deviations of X and Y, respectively.

We may determine that **r** is independent of the choice of units and always lies between −1 and 1 by dividing the previous equation defining **r** by the sample standard deviations, Sx, and Sy.

Now, let's talk about a few more significant characteristics of the linear correlation coefficient, **r**. It's crucial to keep in mind that **r** indicates how strongly two variables are related linearly, and therefore its subsequent qualities only make sense when the data points are dispersed along a line.

The scatterplot's slope is indicated by **r**. When the scatterplot displays a positive slope, the linear correlation coefficient is positive; conversely, when the scatterplot displays a negative slope, it is negative. We refer to definition of **r** and Fig. 2.12, where we have constructed a coordinate system with a second set of axes centered at the location $(\overline{x}, \overline{y})$, to illustrate why this property is valid.

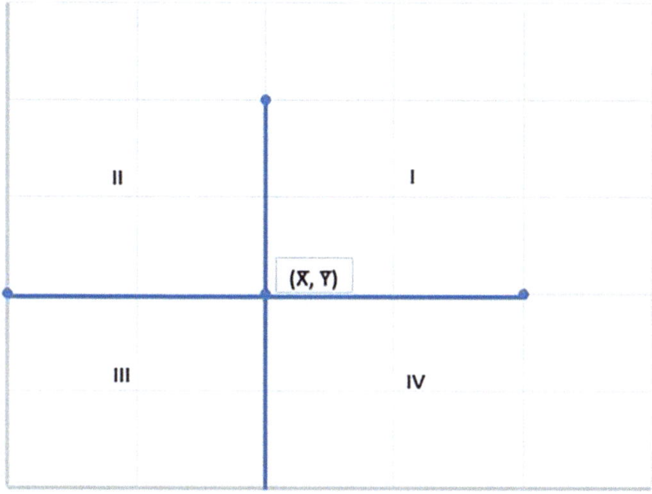

Fig. 2.12 Coordinate system with a second set of axes centered at $(\bar{x}, \bar{y})$

The strength of the linear relation is indicated by the magnitude of **r**. A regression equation useful for making predictions is indicated by a value of **r** near −1 or 1, which suggests a strong linear relation between the variables and the ability of the variable **X** to predict the variable **Y**. A regression equation is either worthless or not very useful for making predictions if the value of **r** is close to zero. This suggests that there is, at most, a weak linear relation between the variables and that the variable **X** is a poor linear predictor of the variable **Y**.

The sign of **r** indicates the kind of linear relation. When **r** is positive, the variables have a positive linear correlation, meaning **Y** tends to increase linearly as **X** increases. This tendency increases as **r** approaches 1. When **r** is negative, it indicates that the variables have a negative linear correlation, which means that **Y** tends to drop linearly as **X** increases. This tendency increases as **r** approaches −1.

Both the sign of **r** and the regression line's slope are the same. When **r** is positive, the regression line slopes upward; conversely, when **r** is negative, the regression line slopes downward.

Figure 2.13 shows an example of strong positive linear correlation.

Figure 2.14 shows an example of perfect positive linear correlation.

Figure 2.15 shows an example of a weak positive linear correlation.

Data points represented in Fig. 2.16 are scattered irregularly and there is almost no linear correlation at all.

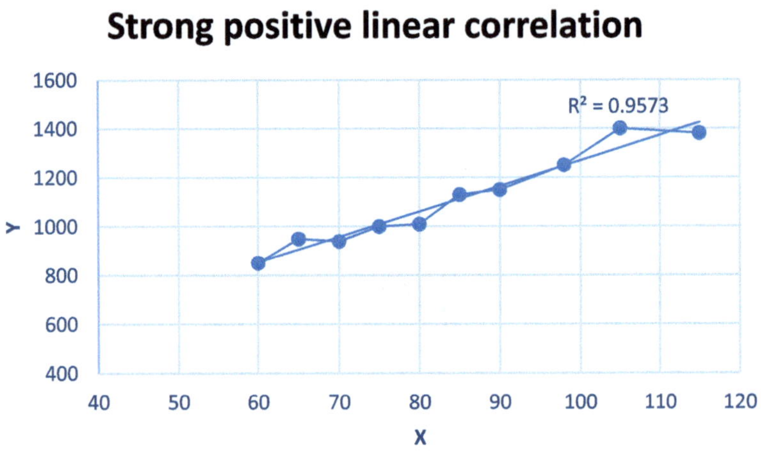

Fig. 2.13 Example of strong positive linear correlation

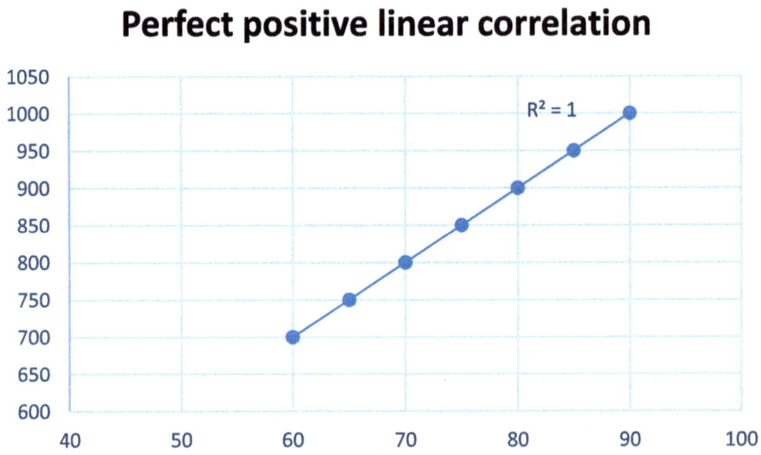

Fig. 2.14 Example of perfect positive linear correlation

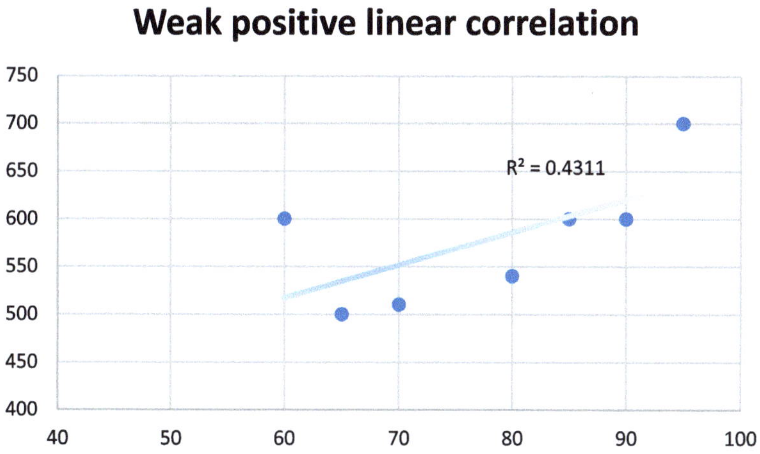

Fig. 2.15 Example of weak positive linear correlation

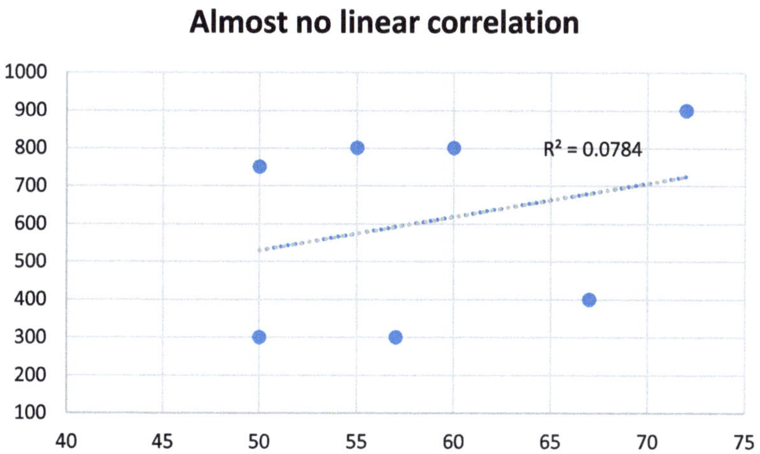

Fig. 2.16 Example of no linear correlation

References

Akoglu, H. (2018). User's guide to correlation coefficients. *Turkish Journal of Emergency Medicine*, *18*(3), 91–93. https://doi.org/10.1016/j.tjem.2018.08.001

Frost, J. (2020). *Introduction to Statistics: An Intuitive Guide for Analyzing Data and Unlocking Discoveries*.

Mukaka, M. M. (2012). Statistics corner: A guide to appropriate use of correlation coefficient in medical research. *Malawi Medical Journal: The Journal of Medical Association of Malawi*, *24*(3), 69–71. http://www.ncbi.nlm.nih.gov/pubmed/23638278

Nikoletseas, M. M. (2020). *Statistics for college students and researchers*.

OpenAI. (2024). *ChatGPT [Large language model]*. https://chat.openai.com

Rovetta, A. (2020). Raiders of the Lost Correlation: A Guide on Using Pearson and Spearman Coefficients to Detect Hidden Correlations in Medical Sciences. *Cureus*. https://doi.org/10.7759/cureus.11794

Saccenti, E., Hendriks, M. H. W. B., & Smilde, A. K. (2020). Corruption of the Pearson correlation coefficient by measurement error and its estimation, bias, and correction under different error models. *Scientific Reports*, *10*(1), 438. https://doi.org/10.1038/s41598-019-57247-4

Weiss, N. A. (2012). *Elementary statistics* (8th ed.). Pearson Education, Inc.

Chapter 3
Hypothesis-Based Versus Hypothesis-Free Approach

3.1 Defining the Research Question

A well-designed research question is a crucial step in the research process. It requires time, expertise, and consideration to ensure the question is well-structured, meaningful, and feasible. Conventional wisdom suggests that it is "good form" to conduct preliminary research in order to get a sense of what has already been studied in the selected research topic, viz., to "comprehend" the different aspects of what one might know. Once the researcher knows the topic well, it may end up being more productive when one is trying to identify gaps in the existing literature and potential areas for exploration.

It is accepted in "evidence-based research" to formulate a proper question, so that the research can be considered appropriate for science. PICO (Population, Intervention, Comparison, and Outcome) is a specialized framework proposed to underly the specialized framework most researchers use to formulate a research question and facilitate a literature review (Eldawlatly et al. 2018). The PICO framework was initially developed for use in medicine and healthcare, and the original context of PICO revolves around clinical questions and interventions (Eriksen and Frandsen 2018; Brown 2020). For a long time, the PICO approach has been considered suitable primarily for clinical studies.

In some cases, the inclusion of time can be a critical factor in the outcome of specific research scenarios; thus, the PICOT framework is also used. The PICOT framework is an extension of the PICO framework, adding the time element to provide even greater clarity when formulating clinical questions.

Recent efforts suggest that the nature of PICO think is more applicable than hitherto thought, that the thinking is more universal in nature, not limited just to clinical work. Several examples demonstrate that the PICO approach can be used in highly heterogeneous studies (Nishikawa-Pacher 2022).

P. Papajorgji, H. Moskowitz, *The Mind of Everyday*,
https://doi.org/10.1007/978-3-031-78078-3_3

Adapting the PICO framework can help social scientists and researchers in other fields frame their research questions, focus their examinations, and design studies, which lead to meaningful insights and deliver evidence-based conclusions.

To show how the PICOT framework in a social study case, let us consider the following: We first lay out the parts of the PICOT framework, with the specifics of the study:

P (Population/Participants): Adolescents
I (Intervention/Exposure): Participation in a comprehensive sex education program
C (Comparison/Context): Participation in a traditional abstinence-only program
(Outcome): Informed decisions about safe sexual practices
T (Timeframe): Within 1 year of program completion

Let us now start at the beginning of the process, when the researcher casts around to find the topic, and works to get the topic into the right form. It should not come as a surprise that "homework" is relevant here, and that little serious research is done without some planning.

When a researcher decides to undertake a study, investing a substantial amount of time in developing research questions, hypotheses, and objectives is essential. It is strongly recommended that the researcher conduct a systematic literature review to increase knowledge and familiarity with the topic and to aid in research development. Current trends and technological advances on the subject during this research phase should be recognized and taken into account when relevant. Depending on the particular research topic, careful information from specialists, mentors, colleagues, and collaborators should be solicited to refine the research question. Addressing this issue with outside help will better shape the research question and guide the research study. A specific, detailed, and well-focused research question is easy to understand. A broad and overloaded research question is challenging to address comprehensively. To summarize, the researcher should assess the constraints related to the selected research question to ensure that it is realistic to conduct the research and solve the problem.

A good research question can be measured or observed and should lead to data that can be collected and analyzed. Addressing a meaningful research question will enrich the existing literature. A fresh perspective on the problem should be provided if the researcher addresses an existing issue.

There is a second approach, abbreviated FINER (Feasible, Interesting, New, Ethical, Relevant) criteria in materializing the research question (Mohanan and Parameswaran 2022). The FINER framework is typically used in research done in the "harder" sciences, in order to assess the feasibility of research projects or research questions. Whereas it is less commonly applied in social sciences than in the harder sciences, the FINER framework may be adapted for social and political research.

Here is an example of how to apply FINER framework to a social studies research question. The research question is: "What is the impact of social media on young adults concerning the Russia-Ukraine war?"

Feasible: Is the research feasible within the available resources and time frame? Can the necessary data and participants be available? Feasibility Assessment: Social media data and population surveys are available. The research can be conducted within the proposed budget and timeline.

Interesting: Is the research question of interest, and is the question likely to contribute to the field of social studies? Interest Assessment: The impact of social media on young adults concerning the war in Ukraine is a relevant and timely topic, with the potential to induce new insights into social studies.

Novel: Does the research question address a gap in the existing literature or provide a new perspective? Novelty Assessment: Whereas there is existing research on this topic and the war is at the center of all media, the proposed study aims to focus specifically on young adults, an underexplored demographic. The study adds a potential novel dimension to the research.

Ethical: Are there ethical considerations in the research, and are they satisfactorily addressed? Ethical Assessment: Ethical considerations, such as informed consent and data privacy, will be carefully addressed in the research design to protect participants' rights. For example, surveys will not include personal data such as the name and the last name.

Relevant: Is the research question relevant to the broader social studies field and society in general? Relevance Assessment: The research question is highly relevant, as it addresses the intersection of technology, youth engagement, and politics, which has significant implications for society.

In this example, the FINER framework helps assess the research question's suitability in social studies. It ensures that the research is feasible, interesting, novel, ethical, and relevant, which are critical considerations for any research project, including those in the social and political sciences.

3.2 Hypotheses and Selecting the Research Approach

Hypotheses are specific predictions or statements that can be tested based on the research question. A well-thought and well-structured research question naturally leads to the formulation of meaningful and testable hypotheses.

One of the first things a researcher must decide while contemplating a new study is the research method to use. There are two main options: a hypothesis-based (Rao 2019; Heun 2018; van Witteloostuijn and van Hugten 2022) and a hypothesis-free approach (Goodman 1999; Yanai and Lercher 2020a). Recently, some authors have suggested that, in some cases, it might be beneficial to use a combined approach, which will provide the advantages of both (Casula et al. 2021; Kitsios and Zintzaras 2009).

For more than two decades, there has been a discussion in the scientific community about the best method to use in scientific studies: the hypothesis-based approach versus the hypothesis-free approach. As for any other issue, different people would have different opinions. Some authors strongly support hypothesis-based research.

Others will criticize the hypothesis-based approach and provide ample arguments for using a hypothesis-free type of research.

3.2.1 The Hypothesis-Based Approach

Many authors have defined what a hypothesis is. Although they use different wording, the meaning is the same. A "hypothesis" is a well-argued analysis of available evidence to provide a realistic (scientific) explanation of existing facts, fill gaps in public understanding of sophisticated processes, and propose a new theory or test (Gasparyan et al. 2019). Although there is a broadly accepted and precise meaning of what a hypothesis is, in some areas, there are different perceptions about the nature of hypotheses, the role of hypotheses in science, and the use of hypotheses within different disciplines (Donovan et al. 2015).

A hypothesis is typically formulated as a statement that suggests a relationship between variables or a potential explanation for a phenomenon. It is a fundamental concept in scientific research and the scientific method. Scientists are taught to frame their experiments with a "hypothesis"—an idea or postulate that must be phrased as a statement of fact so that it can be subjected to falsification (Glass and Hall 2008). In other words, a hypothesis is a proposed explanation or educated guess that can be tested through empirical observation and experimentation.

A hypothesis attempts to answer a specific question. A hypothesis must be measurable and must allow for a response whose validity can be established or rejected. The response may be intermediate to its ultimate objective, and the answer must be essential to achieve the final objective. The hypothesis must be precise and must contain precise statements for which the answer can be confirmed or rejected (Kumar 2014).

It is instructive to consider why a hypothesis may be inadequate (Lund 2022). As stated above, the hypothesis should be formulated clearly and concisely. Thus, a good practice is to avoid coordinating conjunctions like "and" and "or." A hypothesis with coordinating conjunctions is called a composed or compound hypothesis, and they are difficult to verify. In the case of a compound hypothesis, one part can be true, and the other part can be false (Misra et al. 2021).

An example of a complex hypothesis could be: "Going to college increases future income AND makes useful contacts."

The hypothesis comprises two parts:

Increases future income
Makes useful contacts

These two parts are joined by the Boolean AND operator, as shown in Table 3.1. Table 3.1 shows what should be concluded the truth or falsity of the hypothesis. The hypothesis is true only when both composing parts are true. In general, it is advisable not to choose a complex hypothesis and rather avoid them.

Table 3.1 Example of complex hypothesis

A (increases future income)	B (makes useful contacts)	A and B
If this occurs	If this occurs	Then the hypothesis is
True	False	False
False	True	False
True	True	True

The hypothesis should be about the experiment, not about the person experimenting. Thus, it would be wrong to design a hypothesis by wording it as "Person X can do this job." The correct hypothesis should indicate whether this job can or cannot be done independently of different people's abilities (Misra et al. 2021).

Another reason for designing bad hypotheses can be the lack of a solid background knowledge. When the researcher has insufficient knowledge about the subject or problem under investigation, the result may be an uninformed or overly simplistic hypothesis.

The researcher may have personal biases, preconceived notions, or a desire for a particular outcome for the experiment. The preconceived notions can lead to a biased hypothesis, which may not accurately reflect the reality of the situation. Even in cases when a researcher does not have access to all the relevant data or information, the hypothesis may be based on incomplete or inaccurate assumptions, resulting in difficulties when it comes to time to verify or falsify the hypothesis.

A hypothesis should be capable of being proven false. If the hypothesis cannot be tested or falsified through experimentation or observation, then it is not a valid scientific hypothesis. It should be possible to demonstrate that the hypothesis is false if it is indeed false (Lund 2022).

"Lousy" hypothesis: An example of a lousy hypothesis could be the following statement: "Eating fruits is good for you." This hypothesis is problematic for several reasons. First, the hypothesis is vague, lacking specificity. There is no clear definition of the meaning of "good for you." Second, the phrases do not provide any measurable or testable outcomes that could be used to validate the hypothesis. Third, the hypothesis does not offer any criteria or conditions under which it could be proven false. It will be too general to be tested or disproven through experimentation or observation. In contrast, a well-designed hypothesis generally includes variables that can be manipulated and predictions about how changes in those variables will affect the outcome. This above-presented hypothesis lacks any variables or predictions (Chigbu 2019).

Let us look at ways to improve this hypothesis. The hypothesis must be more specific, testable, and include variables and predictions. Consider this improvement: "*Consuming at least five servings of vegetables daily for six months will lead to a statistically significant decrease in cholesterol levels compared to a control group that does not consume vegetables.*" The revised hypothesis clearly defines the variables (vegetable consumption and cholesterol levels); it includes a precise prediction and is testable through experimentation. The data to be collected are the vegetable consumption and the cholesterol level.

The issue of methods for creating solid hypotheses have been dealt with in a number of papers (Omerovic et al. 2010; Kumar 2014; Stojmenović and Milutinovic 2012). There are even specialized software for testing hypothesis (Šašinka et al. 2017).

3.2.2 Testing Hypothesis, Popper's Null-Hypothesis Statistical Testing (NHST)

The history of hypothesis-based research owes a great deal to Karl Popper, a philosopher of science known for his influential work on the philosophy of science, particularly the concept of falsifiability and his ideas regarding the scientific method. Popper's work can be rightly said to have impacted how scientists formulate and test hypotheses.

Popper argued that a hypothesis or scientific theory should be considered scientific if and only if it is potentially falsifiable, meaning that the hypothesis can be tested and proven false through empirical observation or experimentation. According to Popper, a scientific hypothesis should make specific predictions, which could potentially refute the hypothesis if they turn out to be false (Popper 1959).

Popper distinguished between science and nonscience, including pseudoscience. Pseudoscientific claims are typically unfalsifiable because they are formulated in a way that makes it impossible to subject them to empirical testing. In contrast, scientific hypotheses are open to empirical scrutiny and potential falsification (Veronesi 2014).

Popper was critical of the traditional view of science, which relied heavily on inductive reasoning. He argued that induction, which involves generalizing from observed specific instances, was not a reliable method for establishing the truth of scientific theories. Instead, Popper emphasized the importance of deductive reasoning and falsifiability (Pompeu 2018; Vignero and Wenmackers 2021). Popper is famous for introducing the concept of null-hypothesis statistical testing or known as the NHST.

A hypothesis test is a procedure used in statistics to assess whether a particular viewpoint is likely true. Testing procedures follow a strict protocol with the results assumed to be an instance of an underlying "distribution" of similar results. The question is the likelihood of observing the particular results given the measured variation of results. The analysis is done to generate an estimate that the observation was made by random chance, and that the observation is noise. The calculation then generates a "*p*-value," which is the probably of observing the measured outcome if, in fact, the reality was that the data are randomly distributed. It is this type of thinking that undergirds the routine statistical "tests" used in research—*t*-tests, χ^2 tests, Mann–Whitney tests, etc. They are all attempts to determine whether the observed data come from the hypothesis being true, or simply random observations (Walker 2019).

When we want to compare the characteristics of two small, well-defined groups (for example, two groups of children, one group exercising and the other group of non-exercising children), we need to use a summary statistic that can be calculated for each group. Such summaries (means, medians, etc.) form the basis of descriptive statistics and are well-presented in many publications. Usually, both groups are measured for the selected characteristic, and statistics are calculated to allow a conclusion.

There is a problem when we try to compare huge groups or populations: it may be impractical or even impossible to measure everyone in the population, and by the time measurements are finished, the population itself may have varied. To address this problem, random samples are used to estimate values for the entire population. By tradition, the values calculated from samples are referred to as statistics, whereas the unknown population values are called parameters. The field of inferential statistics describes the methods used to estimate population parameters from random samples, how to quantify the level of inaccuracy in a sample statistic, and how to continue using these estimates to compare populations (Colquhoun 2014).

3.2.3 *Example of Hypothesis Verification Using the NHST (Null Hypothesis Statistical Test)*

In order to explain the hypothesis test process, let us assemble a simple example with dummy data in the social sciences. The essence of this example is to use a fictitious dataset to test the hypothesis that participation in a study group improves a students' exam scores.

A professor teaches the same topic, ***The Research Methods***, to two different classes and would like to evaluate whether there are any differences in performances between these two classes regarding assimilating the same materials. For this purpose, the professor designs an experiment and collects some data as follows (Table 3.2):

Research Question How students of different classes perform in the same topic? The answer is in the seven-step process for evaluating the hypothesis.

1. *Define Null and Alternative Hypothesis*

   H(0): $\mu 1 = \mu 2$ thus, students of different classes perform the same.
   H(1): $\mu 1 \neq \mu 2$ thus, there is a difference in performance between classes A and B.

2. **State** α **(alpha).** In the context of hypothesis testing, the symbol ***α*** **(alpha)** represents the **significance level** of a test; it is the probability of rejecting the null hypothesis when it is true, which is known as making a **Type I error**.

Table 3.2 Data collected for the experiment

Class A grades	Class B grades		
3	10		
4	4		
5	5		
7	7		
3	5		
9	5		
8	9	Class A—average grade—$\overline{x1}$	6.85
10	8	Class A—standard deviation (STDA)	2.62
8	10	Number of students = $n1$	20
9	10		
10	5	Class B—average grade = − $\overline{x2}$	7.07
4	7	Class B—standard deviation (STDB)	2.25
4	8	Number of students = $n2$	15
9	9		
5	4		
6			
9			
10			
4			
10			

$$a = 005$$

3. *Calculate the Degrees of Freedom (Degrees of Freedom for Independent Samples t-Test).* In the context of an **Independent Samples t-test**, the term **degrees of freedom (df)** refers to the number of independent pieces of information available to estimate the population variance. It essentially determines the shape of the t-distribution used to calculate the p-value and the critical t-value.

$$\text{df} = (n1-1)+(n2-1)=(20-1)+(15-1)=19+14=33$$

4. *State Decision Rule*

   Figure 3.1 shows the interval of confidence with a = 005. The graph shows that 95% of times the score will fall between the lower and the upper limit. Using the *t*-Table we define the critical value of 20,045 for the degree of freedom 33 and a = 005. Thus, in Fig. 3.1, the lower limit is −20,045 and the upper limit is 20,045. So, when the value of t is outside the interval (−20,045, 20,045) it means that the null hypothesis should be rejected.
5. *Calculate the Test Statistics*

   The formula for calculating t is:

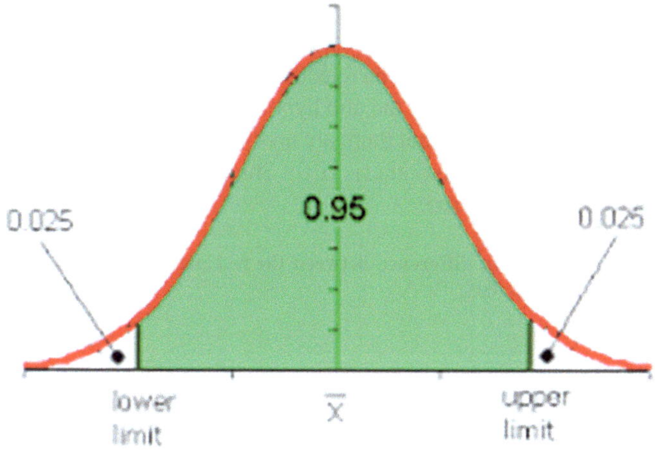

Fig. 3.1 Interval of confidence

$$t = \frac{(\bar{x}1 - \bar{x}2) - (\mu 1 - \mu 2)}{\sqrt[2]{\frac{\mathrm{Sp}^2}{n1} + \frac{\mathrm{Sp}^2}{n2}}}$$

where Sp is the pool variance,

$$\mathrm{Sp}^2 = \frac{\mathrm{SS1} + \mathrm{SS2}}{\mathrm{df1} + \mathrm{df2}}$$

and

$$\mathrm{SS1} = \mathrm{STDEVA}^{2*}\mathrm{df1} = 6.8644^{*}19 = 130.4236$$
$$\mathrm{SS2} = \mathrm{STDEVA}^{2*}\mathrm{df2} = 5.0625^{*}14 = 70.875$$

And

$$\mathrm{Sp}^2 = (130.4236 + 70.875) / (19 + 14) = 201.2986 / 33 = 6.0999$$

Thus,

$$t = \frac{6.85 - 7.06}{\sqrt[2]{\frac{6.0999}{20} + 6.0999 / 15}} = -0.21 / 0.8446 = -\ 0.2486$$

6. *State Results*

   In step 4, we stated that if the value of t is outside the interval (−20,045, 20,045), the null hypothesis will be rejected. Our value of $t = -2486$ is inside this interval, thus we will not reject the null hypothesis, meaning that the null hypothesis is true, and we confirm that there are no differences in performance between classes A and B. Our null hypothesis H(0): Students of different classes perform the same is proven to be true.
7. *State Conclusion*

   There is no significant difference between the test performances of classes A and B with t = − 0.25 and $p > 0.05$.

3.2.4 Example of Using the Effect Size Coefficient

Effect size coefficients enable the researcher and reader understand the practical significance of their findings, by providing more informed interpretations and comparisons across studies. They are exceptionally valuable and informative when combined with confidence intervals, allowing the researcher to estimate a range of plausible values for the effect size.

Consider the average test scores of two groups of students, Group A and Group B, both in the same course. We will use Cohen's d as the **effect size measure to quantify the difference between the groups' means** and calculate the effect size along with its associated confidence interval.

The average score for group A is 75, i.e., the mean of the scores for group A is 75. The average score for group B is 85, i.e., the mean of the scores for group B is 85. The standard deviation for group A is 10, and for group B is 12. We will consider that the sample size for both groups is 50. The confidence level is 95%.

Now for Cohen's *d* and its confidence interval:

Cohen's *d* = (Group B Mean − Group A Mean)/Pooled Standard Deviation
Pooled Standard Deviation = SQRT ((SD_A^2 + SD_B^2)/2)
Cohen's *d* = (85 − 75)/SQRT((10^2 + 12^2)/2) = 10/SQRT(244) ≈ 0.640

Now to calculate the confidence interval for Cohen's *d*:

To calculate the confidence interval for Cohen's *d*, we need the standard error of the estimate (SE). The standard error (SE) of Cohen's d is expressed as:

$$\mathrm{SE} = \mathrm{SQRT}\Big(\big(1/n\_\mathrm{A}\big)+\big(1/n\_\mathrm{B}\big)+\big(d\,\hat{}\,2/\big(2^{*}\big(n\_\mathrm{A}+n\_\mathrm{B}\big)\big)\big)\Big)$$

where *n*_A and *n*_B are the sample sizes of Group A and Group B, respectively. Thus

$$\mathrm{SE} = \mathrm{SQRT}\Big(\big(1/50\big)+\big(1/50\big)+\big(0.640\,\hat{}\,2/\big(2^{*}\big(50+50\big)\big)\big)\Big) \approx 0.107$$

Now, we can use the standard error to calculate the confidence interval. Assuming a t-distribution with degrees of freedom (df) equal to the sum of the sample sizes minus 2:

Margin of Error = $t\_(\text{df}, \alpha/2)$ * SE
Given a 95% confidence level, $\alpha = 0.05$, and using a t-table or calculator, for df = 98 and $\alpha/2 = 0.025$, $t\_(98, 0.025) \approx 1.984$
Margin of Error = $1.984 * 0.107 \approx 0.212$
Confidence Interval: Cohen's $d \pm$ Margin of Error = 0.640 ± 0.212 or (0.428, 0.852).

Let us now interpret the results of our calculation. The difference in test scores between Group A and Group B is estimated to be around 0.64 with a 95% confidence interval ranging from approximately 0.43–0.85. This means that if we were to do this study again and again, say 100 times, we would, of course, have variation, BUT 95% of values of our repeated studies would lie in this region of 0.43–0.85.

If we were to summarize what we just accomplished, it would be like we have a hypothesis, we make the measurements, we calculate the size of the effect and the variability of our measurements, and finally provide a range where we expect 95% of the results to fall (difference between groups) if we were to repeat the study 100 times. The variability that we encounter along with the means from the one study provides us with sufficient information to accept or reject the hypothesis. We also see a deep reliance on statistics to help us accept or reject what we believe to be the case.

3.2.5 The Hypothesis-Free Approach

The hypothesis-free approach, also known as data-driven research or exploratory data analysis, is a relatively new scientific method, which differs from the traditional hypothesis-driven approach. The hypothesis-free approach became popular with new technologies, such as big data and advanced computational techniques, which allow for exploring large datasets without a predefined hypothesis.

The hypothesis-free approach involves analyzing data without starting with a specific, predefined hypothesis. Instead of testing a predefined hypothesis, researchers explore the data to identify correlations, patterns, and insights, which may lead to the formulation of new hypotheses. This approach is common in those domains where the underlying mechanisms are poorly understood or where the data is too complex to allow for a simple hypothesis to be formulated at the start of the research effort.

The hypothesis-free approach is data-driven. There are a few essential steps to prepare the data for analysis.

First, a large amount of data and comprehensive datasets should be collected from various sources. This step is crucial and often involves technologies designed for "big data."

Second, he collected data must be preprocessed to ensure it is clean, organized, and suitable for complex analysis. This step might involve handling missing values, removing outliers, and standardizing formats.

Third, the research invokes data exploration, which is performed using statistical and computational tools to explore the data. The exploration process can include visualization techniques, descriptive statistics, and clustering algorithms. Data exploration attempts to identify patterns, correlations, or anomalies within the data. Data explorations stage often uses machine learning algorithms, especially those falling into the category of "unsupervised learning."

Fourth, the vast datasets can help formulate new hypotheses based on the insights gained from the preparatory stages of data exploration and pattern recognition. Once new hypotheses are generated, they can be tested using traditional hypothesis-driven approaches.

Hypothesis-free approaches can uncover relationships and patterns not previously known. This approach allows researchers to explore data without being confined to a specific hypothesis. Hypothesis-free approaches are well-suited to handle large datasets where traditional hypothesis-driven approaches may not be feasible.

Genomics and bioinformatics use hypothesis-free approaches in order to identify genetic markers and pathways involved in diseases. An excellent example is "The Human Genome Project" (Collins and Fink 1995), where researchers analyzed vast amounts of genetic data without a specific hypothesis, showing groundbreaking discoveries in genetics.

Finance also uses hypothesis-free approaches in order to detect patterns in market data to predict stock movements. Retailers also use hypothesis-free methods. A typical example is the "Market Basket Analysis" to analyze transaction data and identify customers purchasing patterns, which can inform marketing and inventory decisions (Milutinovic and Salom 2016). Mind Genomics, a typical hypothesis-free analysis tool, is successfully used to study market segmentation (Moskowitz and Gofman 2007; Zemel et al. 2018).

And the list continues. Two radically different examples finish the list. Environmental sciences use hypothesis-free approaches to discover climate patterns and their effects on ecosystems (Gere et al. 2010). Political scientists use hypothesis-free approaches to analyze social media to understand user behavior and trends (Kover et al. 2022).

3.2.6 Comparison of the Two Approaches

Hypothesis-based research is a classic approach to scientific exploration largely used by the research community worldwide. This approach involves forming a straightforward prediction or hypothesis about a phenomenon under study and then performing experiments or collecting data in order to verify the hypothesis. Hypothesis-based research has been widely used in many fields of science, including biology, physics, psychology, and social sciences (Brand 1999; Dünki and Dressel 2011; Heun 2018) to name a few.

Hypothesis-based research has been and remains a highly regarded and widely used approach in many areas of science. It is still considered the gold standard for scientific investigation, providing a rigorous and systematic pattern to verify ideas and theories and help our understanding of the world (Wiley 2008; Biesecker 2013; Shardlow et al. 2018).

Opponents of hypothesis-based research are supporters of the hypothesis-free approach to scientific research, known as exploratory data analysis or discovery science (Aebersold et al. 2000). Hypothesis-free research searches for patterns and relationships in data without preconceived ideas or hypotheses about what the data might reveal (Goodman 1999; Yanai and Lercher 2020a). In this approach, researchers use various techniques to analyze large datasets and identify patterns and relationships that may not have been predicted.

The "battle" for the proper scientific approach has been quite public, occasionally strident, present among scholars of different take on this issue. The highest level of debate has been between the famous journals Nature and The Science. Nature Biotechnology published "Equipping Scientists for the New Biology" (Aebersold et al. 2000), and The Science published "Hypothesis-free? No such thing" (Wiley 2008). Hypothesis-based supporters argue that it is simply unrealistic to expect preliminary studies to lead to significant biological insights without a hypothesis. The hypothesis-based researchers further acknowledge that systems biology *MAY* be the future of biology, but researchers still need hypotheses to take us where we want to go. To these hypothesis-based researchers, the big fear is that hypothesis-free approach will create "bunches of data," whose interpretation is ambiguous and, in the end, meaningless and useless.

A retort from opponents of the hypothesis-based approach argues that small, autonomous, highly specialized, and hypothesis-driven groups have ended up carrying out university-based research programs with out-of-date thinking, small samples sizes, convenience samples of university studies. The argument is that these small, isolated groups have access to limited ranges of topics, limited ideas, and produce slow-moving researcher. This structure has rendered the typical research group unable to react quickly to significant technological advances or changes in scientific thinking (Aebersold et al. 2000).

One of the main criticisms of the hypothesis-based approach is that the mantra of "start with a hypothesis" might limit a researcher's perspective (Yanai and Lercher 2020a). When a researcher is too attached to the hypothesis, such attachment could be blinding, ending up with the otherwise well-meaning researcher unintentionally ignoring data not supporting the hypothesis, or worse, seeking to diminish cognitive dissonance by conjuring up alternative explanations for their observations. In Hesse's words, put in Siddhartha's mouth "*When someone seeks, it can easily happen, but his eyes only see the thing he is seeking and that he is incapable of finding anything, incapable of taking anything in because he is always only thinking about what he is seeking, because he has an object, a goal because the goal possesses him*" (Hesse 1922).

The opponents of the hypothesis-based approach use the so-called "The gorilla experiment" (Simons and Chabris 2010). This experiment (authors) shows two groups of people passing the ball to each other, and the audience is asked to count the number of times they pass the ball. In the middle of the experiment, a person dressed as a gorilla appears in the middle of the court. The experiment demonstrates the small number of people recognizing the gorilla in the court, as they are focused on counting the ball passing from one team to the other. This experiment became famous, showing how easy it is to get deceived by our own intuition (Simons and Chabris 2010).

Authors of "The Invisible Gorilla," argue that it is a mistake to see this demonstration as proof that some people are easily duped. Instead, they demonstrate that this is an inherent limitation of cognition. They further state that the real problem is "the illusion of attention"—that people are often unaware of these limitations. People think that they see the world as it really is, but "our vivid visual experience belies a striking mental blindness." Humans overrate their cognitive abilities. "Intuition deceives" (Holderness 2014).

Researchers collect data with some fixed goal in mind, testing a specific hypothesis. However, researchers are likely to miss other exciting points of view where their mind is switched to hypothesis testing. To remedy this, researchers must adopt a different mind-set—the one of exploration. Like an explorer lost in the middle of the jungle, looking left and right to find the path reaching the top of the hill, researchers must look at the data from as many angles as possible. The explorer mind-set allows playfulness with the data, comparing everything to everything else until the right path is discovered (Yanai and Lercher 2020a).

Another argument the anti-hypothesis-based scientists provide that goes with Siddhartha's saying is that the history of science demonstrates that many important discoveries have been achieved by pure chance and without any predefined hypothesis. The most famous case of serendipity is the discovery of penicillin. Fleming noticed that a mold called *Penicillium notatum* had killed bacteria in a petri dish. He was not explicitly searching for antibacterial agents but realized the potential of the mold. This discovery led to the development of penicillin, the world's first antibiotic. Another famous serendipity case is the discovery of X-rays. Röntgen was working with cathode-ray tubes when he noticed a fluorescent effect on a nearby screen. Investigating further, he discovered X-rays, which revolutionized medicine. History has many other serendipitous discoveries, such as microwaves, vulcanized rubber, etc. By focusing only on the hypothesis-driven approach, the scientific community might undervalue exploratory research, which could lead to novel discoveries. Testing a hypothesis based on an existing paradigm continuously may be a distraction from new theories.

A hypothesis-free data analysis is almost five times more likely to make an exciting observation than a hypothesis-focused analysis (Yanai and Lercher 2020b). It is advisable to teach students the benefits of visualization. Answering the specific hypothesis-driven questions did not require plotting the data. Students driven by specific hypotheses often skipped this simple step toward a broader data exploration. Students' work links the psychological effect of selective attention and hypothesis-driven data analysis. The results of small-scale experiments provide evidence that, indeed, there is a hidden cost to having a hypothesis (Yanai and Lercher 2020b).

3.3 All Research Ends Up Being "Hypothesis-Light"

There is a fever for publishing in academia and many research environments; the famous "publish or perish" slogan (Harzing 2010) can push researchers to produce positive findings, potentially leading to questionable research practices (Saxena et al. 2023). Hypothesis testing often relies on quantitative data and statistical methods. Opponents argue that this can downplay the significance of qualitative insights, narratives, and subjective experiences, which might offer more profound or alternative understandings of a phenomenon. The pressure to publish often pushes authors to lower the publication quality (Sarewitz 2016).

Also, it happens that there is a gap between the defined research questions and the formulated hypothesis; they may not converge. Not all research questions or fields lend themselves well to hypothesis testing (Farrugia et al. 2010). A divergence between a research question and a hypothesis occurs when the research question asks about a general relationship or phenomenon. In contrast, the hypothesis offers a specific, testable prediction about that relationship or phenomenon.

Let us consider an example in the social sciences field to shed more light on this issue. Let us suppose that **social media and Self-Esteem** is the topic under investigation. The research question is to determine whether there is a relation between the number of hours spent on social media and an individual's self-esteem. Let us suppose that the hypothesis is ***The more hours individuals spend on social media per day, the lower their self-esteem will be***.

Whereas the research question has a broader focus, the potential connection between social media and self-esteem, the hypothesis has a narrow focus, only whether the increasing number of hours spent on social media negatively affects self-esteem.

Another example could be trying to form a hypothesis in some exploratory or descriptive studies might be premature or inappropriate. Furthermore, focusing on a single hypothesis can sometimes lead to oversimplified interpretations of complex phenomena, especially if nuanced or multifaceted factors are at play (Lund 2022).

To finalize this discrepancy between the research question and the hypothesis, the researcher must consider that whereas research questions tend to be more open-ended and exploratory, hypotheses make precise predictions and are often designed to be validated or invalidated through experimentation or analysis.

The hypothesis-free approach's opponents state that finding meaningful relationships in complex datasets requires starting with the appropriate data. A hypothesis usually takes the form of a mechanistic relationship between a specific cause and a consequent effect. The cause–effect relationship will almost always depend on the experimental context. There are some circumstances when data must be gathered without context or hypothesis to characterize a system. However, expecting such preliminary studies to lead to significant biological insights is unrealistic. For this, a hypothesis is needed (Wiley 2008).

Observation and data are always based on some hypothesis or theory. Data is meaningless without some form of hypothesis or theory. Any data exploration, regardless of the level of informality, is necessarily guided by some form of expectations. Even informal suspicions or assumptions are types of proto-hypothesis (Felin et al. 2021).

References

Aebersold, Ruedi, Leroy E. Hood, and Julian D. Watts. 2000. "Equipping Scientists for the New Biology." *Nature Biotechnology* 18(4): 359–359. https://www.nature.com/articles/nbt0400_359.

Biesecker, L. G. 2013. "Hypothesis-Generating Research and Predictive Medicine." *Genome Research* 23(7): 1051–53. http://genome.cshlp.org/cgi/doi/10.1101/gr.157826.113.

Brand, Richard. 1999. "Hypothesis Based Research The Maturation of Orthopaedic Science." *Iowa Orthopedic Journal* 19: 11–17.

Brown, David. 2020. "A Review of the PubMed PICO Tool: Using Evidence-Based Practice in Health Education." *Health Promotion Practice* 21(4): 496–98. http://journals.sagepub.com/doi/10.1177/1524839919893361.

Casula, Mattia, Nandhini Rangarajan, and Patricia Shields. 2021. "The Potential of Working Hypotheses for Deductive Exploratory Research." *Quality & Quantity* 55(5): 1703–25. https://link.springer.com/10.1007/s11135-020-01072-9.

Chigbu, Uchendu. 2019. "Visually Hypothesising in Scientific Paper Writing: Confirming and Refuting Qualitative Research Hypotheses Using Diagrams." *Publications* 7(1): 22. https://www.mdpi.com/2304-6775/7/1/22.

Collins, Francis S, and Leslie Fink. 1995. "The Human Genome Project." *Alcohol health and research world* 19(3): 190–95. http://www.ncbi.nlm.nih.gov/pubmed/31798046.

Colquhoun, David. 2014. "An Investigation of the False Discovery Rate and the Misinterpretation of p -Values." *Royal Society Open Science* 1(3): 140216. https://royalsocietypublishing.org/doi/10.1098/rsos.140216.

Donovan, Shannon M., Michael O'Rourke, and Chris Looney. 2015. "Your Hypothesis or Mine? Terminological and Conceptual Variation Across Disciplines." *SAGE Open* 5(2): 215824401558623. http://journals.sagepub.com/doi/10.1177/2158244015586237.

Dünki, R. M., and M. Dressel. 2011. "F-Ratio Test and Hypothesis Weighting: A Methodology to Optimize Feature Vector Size." *Journal of Biophysics* 2011: 1–11. https://www.hindawi.com/journals/jbp/2011/290617/.

Eldawlatly, Abdelazeem et al. 2018. "Appearance of Population, Intervention, Comparison, and Outcome as Research Question in the Title of Articles of Three Different Anesthesia Journals: A Pilot Study." *Saudi Journal of Anaesthesia* 12(2): 283. https://journals.lww.com/10.4103/sja.SJA_767_17.

Eriksen, Mette Brandt, and Tove Faber Frandsen. 2018. "The Impact of Patient, Intervention, Comparison, Outcome (PICO) as a Search Strategy Tool on Literature Search Quality: A Systematic Review." *Journal of the Medical Library Association* 106(4). http://jmla.pitt.edu/ojs/jmla/article/view/345.

Farrugia, P., B. A. Petrisor, F. Farrokhyar, and M. Bhandari. 2010. "Research Questions, Hypotheses and Objectives." *Canadian Journal of Surgery* 53(4): 278–281.

Felin, Teppo et al. 2021. "The Data-Hypothesis Relationship." *Genome Biology* 22(1): 57. https://genomebiology.biomedcentral.com/articles/10.1186/s13059-021-02276-4.
Gasparyan, Armen Yuri et al. 2019. "Scientific Hypotheses: Writing, Promoting, and Predicting Implications." *Journal of Korean Medical Science* 34(45). https://jkms.org/DOIx.php?id=10.3346/jkms.2019.34.e300.
Gere, A. et al. 2010. "Public Driven and Public Perceptible Innovation of Environmental Sector." In *Innovation Strategies in Environmental Science*, ed. C Galanakis. Elsevier, 69–106.
Glass, David J., and Ned Hall. 2008. "A Brief History of the Hypothesis." *Cell* 134(3): 378–81. https://linkinghub.elsevier.com/retrieve/pii/S0092867408009537.
Goodman, Laurie. 1999. "Hypothesis-Limited Research." *Genome Research* 9: 673–674.
Harzing, A.W. 2010. *The Publish or Perish Book.*
Hesse, Herman. 1922. *Siddhartha.* Dover Thri.
Heun, Reinhard. 2018. "How to Write a Scientific Paper: A Hypothesis-Based Approach." *Global Psychiatry* 1(1): 3–6. https://www.sciendo.com/article/10.2478/gp-2018-0004.
Holderness, Graham. 2014. "Review Article: Shakespeare and Perception." *Critical Survey* 26(3). http://berghahnjournals.com/view/journals/critical-survey/26/3/cs260311.xml.
Kitsios, Georgios D., and Elias Zintzaras. 2009. "Genome-Wide Association Studies: Hypothesis-'Free' or 'Engaged'?" *Translational Research* 154(4): 161–64. https://linkinghub.elsevier.com/retrieve/pii/S193152440900214X.
Kover, A., P. Papajorgji, and H. Moskowitz. 2022. *Applying Mind Genomics to Social Sciences.* IGI Global. https://services.igi-global.com/resolvedoi/resolve.aspx?doi=10.4018/978-1-7998-8409-5.
Kumar, R. 2014. *Research Methodology A Step-by-Step Guide for Beginners.* Sage Publications.
Lund, Thorleif. 2022. "Research Problems and Hypotheses in Empirical Research." *Scandinavian Journal of Educational Research* 66(7): 1183–93. https://www.tandfonline.com/doi/full/10.1080/00313831.2021.1982765.
Milutinovic, Veljko, and Jakob Salom. 2016. *Mind Genomics: A Guide to Data-Driven Marketing Strategy (SpringerBriefs in Business).* 1st Edition. eds. V. Milutinovic and Salom J. Springer.
Misra, Durga Prasanna et al. 2021. "Formulating Hypotheses for Different Study Designs." *Journal of Korean Medical Science* 36(50). https://jkms.org/DOIx.php?id=10.3346/jkms.2021.36.e338.
Mohanan, Saritha, and Narayanan Parameswaran. 2022. "FINER Criteria—What Does It Mean?" *Cosmoderma* 2: 115. https://cosmoderma.org/finer-criteria-what-does-it-mean/.
Moskowitz, H., and A. Gofman. 2007. *Selling Blue Elephants: How to Make Great Products That People Want before They Even Know They Want Them.* Pearson Education.
Nishikawa-Pacher, Andreas. 2022. "Research Questions with PICO: A Universal Mnemonic." *Publications* 10(3): 21. https://www.mdpi.com/2304-6775/10/3/21.
Omerovic, S., S. Tomazic, M. Milutinovic, and V. Milutinovic. 2010. "Methodology for Written and Oral Presentation of Research Results." *Journal of Professional Issues in Engineering Education and Practice.*
Pompeu, Fernando A M S. 2018. "A Little Bit More about Popper's Philosophy." *BMJ Open Sport & Exercise Medicine* 4(1): e000401. https://bmjopensem.bmj.com/lookup/doi/10.1136/bmjsem-2018-000401.
Popper, K. 1959. *The Logic of Scientific Discovery.* Routledge.
Rao, UmadeviKrishnamohan. 2019. "Hypothesis-Driven Research." *Journal of Oral and Maxillofacial Pathology* 23(2): 168. http://www.jomfp.in/text.asp?2019/23/2/168/264803.
Sarewitz, Daniel. 2016. "The Pressure to Publish Pushes down Quality." *Nature* 533(7602): 147–147. https://www.nature.com/articles/533147a.
Šašinka, Čeněk, Kamil Morong, and Zdeněk Stachoň. 2017. "The Hypothesis Platform: An Online Tool for Experimental Research into Work with Maps and Behavior in Electronic Environments." *ISPRS International Journal of Geo-Information* 6(12): 407. http://www.mdpi.com/2220-9964/6/12/407.
Saxena, Sangeeta, Thomas Godfrey, and John Yannessa. 2023. "Publish or Perish: Time for a Rethink?" *Qeios.* https://www.qeios.com/read/FPZJKP.

Shardlow, Matthew et al. 2018. "Identification of Research Hypotheses and New Knowledge from Scientific Literature." *BMC Medical Informatics and Decision Making* 46(18).

Simons, D.J., and C. F. Chabris. 2010. *The Invisible Gorilla: And Other Ways Our Intuitions Deceive Us*.

Stojmenović, I., and V Milutinovic. 2012. "How to Write Research Articles in Computing and Engineering Disciplines." *Journal of Applied Sciences* 9(1): 42–50.

Veronesi, Carlo. 2014. "Falsifications and Scientific Progress: Popper as Sceptical Optimist." *Lettera Matematica* 1(4): 179–84. http://link.springer.com/10.1007/s40329-014-0031-7.

Vignero, Leander, and Sylvia Wenmackers. 2021. "Degrees of Riskiness, Falsifiability, and Truthlikeness." *Synthese* 199(3–4): 11729–64. https://link.springer.com/10.1007/s11229-021-03310-5.

Walker, J. 2019. "Hypothesis Tests." *BJA Education* 19(7): 227–31.

Wiley, H.S. 2008. "Hypothesis-Free? No Such Thing." *The Scientist* 5(31).

van Witteloostuijn, Arjen, and Joeri van Hugten. 2022. "The State of the Art of Hypothesis Testing in the Social Sciences." *Social Sciences & Humanities Open* 6(1): 100314. https://linkinghub.elsevier.com/retrieve/pii/S2590291122000687.

Yanai, Itai, and Martin Lercher. 2020a. "A Hypothesis Is a Liability." *Genome Biology* 21(1): 231. https://genomebiology.biomedcentral.com/articles/10.1186/s13059-020-02133-w.

Yanai, I., and M. Lercher. 2020b. "Selective Attention in Hypothesis-Driven Data Analysis." *bioRxiv preprint*.

Zemel, R. et al. 2018. "Uncovering Consumer Mindsets Regarding Raw Beverages." *Food and Nutrition Sciences* 9(3): 259–67.

Chapter 4
Qualitative-Based Versus Quantitative-Based Research

4.1 Epistemology and How the Research Community Operates Within It

Since the early days of civilization, people have wondered how they know, what is considered knowledge, and how knowledge increases (Cunningham and Fitzgerald 1995). These issues related to how knowledge is acquired, justified, and organized are the domain of epistemology (Hetherington 2019). Epistemology is the branch of philosophy that deals with the nature, scope, and limits of human knowledge.

Epistemology explores questions related to how knowledge is acquired, justified, and organized, seeking to understand the various ways in which we come to know and understand the world around us. Epistemology examines the processes, methods, and criteria by which we distinguish between true and false beliefs, exploring concepts such as belief, justification, evidence, skepticism, and certainty. In essence, epistemology deals with understanding the nature of knowledge itself and the ways in which we can acquire and validate it (Wang 2020).

The research community uses two main approaches to understand and explain phenomena surrounding everyday life; the qualitative and quantitative approaches, respectively. For quite some time, there has been a vivid debate in the scientific community regarding the relative virtues of the quantitative and qualitative "methods" and worldviews, followed by questions such as which methods are "better," or more reasonably, what are the conditions that favor using one of the two classes of measuring, and why.

It is generally by now recognized that the exact definition of "qualitative" and "quantitative" differ from author to author. Indeed, not surprisingly, there seems to be greater consensus on what makes them different and what to expect from each of them. When it comes to the "meaning" of data and results, however, epistemology rears its head, raising the possibility that some of the differences are just the ordinary variation in answers developed by different methods of the same type or

P. Papajorgji, H. Moskowitz, *The Mind of Everyday*,
https://doi.org/10.1007/978-3-031-78078-3_4

whether the issue is deeper, a result of difference in epistemological considerations (Bryman 1984).

Researchers must address a number of important issues when starting to plan for a research study. The first issue is whether a qualitative or quantitative approach best fits the study. Another crucial point is presenting the results in a journal paper or a dissertation. Thus, researchers need to handle relevant design decisions such as conceiving a paradigm, articulating a purpose for the study, identifying the research questions and hypotheses, using theory, and defining and communicating the significance of the study (Creswell 1994). It is challenging and often unproductive to state that one technique/approach has absolute set of advantages over the others. One technique may generate one type of information, and another technique may generate a different type of information The problem under investigation properly dictates the methods of investigation (Trow 1957).

Several researchers have replaced the term quantitative with synonyms such as positivist and empiricist. The term quantitative has been replaced with "naturalistic" field research, "ethnographic," "interpretivist," and "constructivist" (Bryman 1984).

In summary, whereas the principles of quantitative research can be traced back to ancient times, the formal development of qualitative research methods occurred later. The development was promoted in part by the demand for greater in-depth, contextually rich approaches to the study of human behavior and social phenomena. Over time, both quantitative and qualitative research methods have evolved to the point where they now coexist as essential components of the social and scientific research landscape.

4.2 Quantitative Research Methods

Quantitative research measures and analyzes numerical data to identify patterns, relations, and generalizable trends, often emphasizing objective measurements and statistical analysis. Researchers collect numerical data through surveys, experiments, and structured observations. They often use statistical data analysis techniques, such as regression analysis or hypothesis testing. Quantitative research uses a more extensive data sample size in order to establish the statistical robustness and the generalizability of the findings.

Quantitative research findings are presented using statistical measures, such as means, percentages, correlations, or effect sizes. They allow for generalizing to larger populations and drawing conclusions based on statistical evidence. Quantitative research provides precise and replicable data, facilitates hypothesis testing and theory building, allows for statistical comparisons and generalizability, and can provide evidence for cause–effect relationships.

Quantitative methods examine the effects of specified circumstances (referred to as an independent variable) on an outcome of interest (referred to as a dependent variable). In this defined context, causal inferences are drawn from direct observation, as in true experiments, or associations established through statistical analysis.

These methods are most effective when the content is constrained or controlled that the study events are free from any undefined influence. Under these circumstances, reproducibility is high, and the results are likely to reliably predict the outcome of the same event in the future.

Quantitative research is generally considered to be older than qualitative research. Quantitative research has its roots in the scientific method and the work of early natural philosophers and mathematicians, going far back to ancient civilizations. The development of quantitative research can be traced through milestones in history. Thus, ancient Greece and Babylonian scholars (around 2000–500 BCE) contributed to quantitative methods, including developing mathematical principles and numerical systems.

Gillings (2003) provides insights into the mathematical achievements of ancient civilizations, including the Egyptians, Babylonians, and Greeks, respectively, exploring numerical systems, mathematical notations, and quantitative approaches used by these ancient cultures. Boyer and Merzbach (2011) provide a comprehensive history of mathematics, covers the contributions of various ancient civilizations, including the Greeks and Babylonians, delving into their mathematical methods, discoveries, and quantitative techniques. Imhausen et al. (2007) present a book that includes translated mathematical texts from ancient civilizations, including the Greeks and Babylonians, providing primary sources which offer insights into their quantitative research methods and mathematical knowledge.

Wolpert (2002) shows Greek science was encouraged and restrained by wholly "unscientific" cultural interests, ideas, and biases arising from the day's language and paradigms. The book provides an exciting example regarding the prediction of eclipses. The book states that eclipses were not an issue of concern for ancient astronomers until the day some "non-scientific" authors, such as the historian Livy, embellishing a good story with a moral, suggested that it should be.

One of the most representative book describing Greek's contribution to mathematics and quantitative methods is "A History of Greek Mathematics" by Sir Thomas Heath (1981).

Of all the books written about the Greek genius, none is more impressive and even awe-inspiring than that revealed by Thomas Heath's book "A History of Greek Mathematics." Heath presents that history with the scholarly comprehension and comprehensiveness that makes his work as obviously classic now as when it first appeared in 1921. The linkage and unity of mathematics and philosophy is the backbone of the outline for the entire history. Heath covers in sequence Greek numerical notation, Pythagorean arithmetic, Thales and Pythagorean geometry, Zeno, Plato, Euclid, Aristarchus, Archimedes, Apollonius, Hipparchus and trigonometry, Ptolemy, Heron, Pappus, Diophantus of Alexandria, and the algebra. Interspersed are sections devoted to the history and analysis of famous problems: squaring the circle, angle trisection, duplication of the cube, and an appendix on Archimedes' proof of the subtangent property of a spiral (Heath 1981).

Another well-known Greek mathematician of these times is Diophantus, who lived in Alexandria, Egypt, from ca. 201 to 285. Diophantus became famous for his series of books entitled *Arithmetica. In his books, he* presents algebra in a way we would recognize today; he was the first mathematician to do so. Diophantus used symbols for variables, positive and negative numbers, and fractions at the time were very advanced. Many of the concepts he presented in Arithmetica are still used in modern mathematics. Diophantus' ideas paved the way for future mathematical advances going forward centuries later.

In his work, *Arithmetica* Diophantus addresses expressing and solving equations to find one or more unknowns. Diophantus used symbols for the unknown (like "x"), something not seen before in mathematical writings. Almost after ten centuries, *Arithmetica* was written; the concepts presented in this book influenced notable mathematicians like Pierre de Fermat, a French scholar from the seventeenth century. Fermat is considered the founder of modern number theory. Diophantus's work largely inspired him, and he wrote Fermat's Last Theorem in the margin of a 1621 French translation of *Arithmetica* (Langham 2022).

The Greek and Babylonian civilizations contributed fundamentally to sciences, especially philosophy, law, and mathematics; the modern world owes them much gratitude.

It is interesting to note that Rome, although powerful, was able to conquer half of the world; however, it did not make a significant contribution to theoretical sciences. There is a well-known saying that Rome militarily occupied Greece, but culturally Greece occupied Rome. Many Greek philosophers may have been enslaved people in Rome but ended up "occupying" Rome with the power of their minds.

Rome's mathematicians made significant contributions to quantitative methods, particularly in the context of practical applications and engineering. Whereas Roman mathematics was not as theoretical or advanced as Greek or Babylonian mathematics, it was crucial to their engineering, architecture, and administrative systems (Kessener 2021). Thus, it is essential to acknowledge that due to the limited surviving texts specifically dedicated to Roman mathematics, the references may provide a broader context of Roman achievements and practices rather than detailed accounts of individual mathematicians.

During the Islamic Golden Age (eighth to fourteenth century CE), scholars like Al-Kindi and Al-Khwarizmi significantly advanced algebra and numerical analysis. Al-Kindi, also known as Abu Yūsuf Yaʿqūb ibn ʾIsḥāq al-Kindī, was a prominent Arab philosopher, scientist, and mathematician during the Islamic Golden Age. He contributed to many fields, including mathematics, astronomy, medicine, and philosophy (Adamson 2008; Urvoy 2017; Viegas 2006).

Around the year 820, Al-Khwarizmi (his full name is Muhammad ibn Musa al-Khwarizmi) wrote the book Algebra (Kitab al-Jabr wa-al-muqabala), the first scientific text in history to systematically present algebra as a mathematical discipline independent of geometry and arithmetic (Rashed 2010). Al-Khwarizmi is arguably the most influential mathematician of the Middle Ages. He developed two distinct branches of mathematics, both of which owe their name to him: algebra and algorithms. He introduced Hindu-Arabic numerals and algebra to European mathematicians, as well as significant contributions to geography and astronomy as well (Brezina 2005).

The twelfth-century philosopher Ibn Rushd, also known as Averroes, played a critical role in disseminating classical philosophy to Islam. Ibn Rushd's work profoundly influenced Western scholasticism and aspects of Renaissance thought. Ibn Rushd maintained that the most profound truths must be approached through rational analysis and that philosophy could lead to the final truth. He tried to reconcile religion with philosophy without amalgamating or eradicating their differences (Urvoy 2017).

The Scientific Revolution of the sixteenth and seventeenth centuries, led by scientists like Galileo Galilei and Johannes Kepler, is considered the birth of modern quantitative research. Galilei and Kepler brought a more systematic and empirical approach to scientific inquiry, underscoring the crucial role of measurement and quantitative analysis.

Galileo Galilei ***pioneered the experimental scientific method*** and was the first to use a refracting telescope to make important astronomical discoveries. Galileo also comments critically on scientific methodology in response to challenges to his scientific ideas. He ultimately writes this fascinating treatise called *The Assayer* that questions many of the presuppositions about how we know and what we know. Galileo is one of the first scientists to reflect on epistemology. Galileo is the most critical person trying to define the relationship between science and religion. His ***Letter to the Grand Duchess*** *Christina is* a foundational document for people trying to understand the relationship between these two essential ways of knowing and believing (Fox 2016).

Johannes Kepler is one of the most significant representatives of the so-called sixteenth and seventeenth century Scientific Revolution. In his work, it is possible to identify two main tendencies. The first links to Platonism and makes geometry play a central role in the structure of the world. The second relates to the Aristotelian tradition, accentuating the role of experience and causality in epistemology (Di Liscia 2021).

During the eighteenth and nineteenth centuries, the development and formalization of the quantitative approach in various fields, particularly in the natural and social sciences, were influenced by several notable figures. Some representatives of the quantitative approach during this period include:

Sir Francis Galton (1822–1911): Galton was a polymath and a pioneer in applying statistical methods to various scientific disciplines. He is known for his work in eugenics, meteorology, and the study of heredity. Galton developed statistical techniques such as correlation and regression analysis to study the relationship between variables (Gillham 2001).

Adolphe Quetelet (1796–1874): Quetelet was a Belgian mathematician, statistician, and sociologist. He introduced the concept of the "average man" and conducted extensive statistical analyses of social and demographic data. He played a significant role in promoting the use of statistical methods in social sciences (Eknoyan 2007).

Carl Friedrich Gauss (1777–1855) was a German mathematician, astronomer, and physicist whose profound contributions to various fields of science have had a lasting impact on modern knowledge. Gauss is often referred to as the "Prince of

Mathematicians" due to his remarkable insights and groundbreaking discoveries. His legacy encompasses a wide range of disciplines, from pure mathematics to applied sciences. Gauss made foundational contributions to numerous areas of mathematics, revolutionizing the field with his innovative insights and techniques. He is perhaps best known for: Number Theory, Non-Euclidian Geometry, Statistics, Astronomy and Physics, Celestial Mechanics, Magnetism, and Gauss's Law.

Gauss's work profoundly influenced subsequent generations of mathematicians, scientists, and researchers. His methods and insights laid the groundwork for much of modern mathematics and physics, shaping how we understand the natural world. Gauss's contributions extended beyond individual theorems and equations; he established a rigorous and systematic approach to mathematical and scientific inquiry that continues to inspire and guide researchers to this day (Tent 2006).

William Stanley Jevons (1835–1882): Jevons was an English economist and logician who contributed to the development of the marginal utility theory and the application of mathematical and statistical methods in economics. He emphasized the importance of quantitative analysis in economic research (Mosselmans 2015).

Auguste Comte (1798–1857): Comte was a French philosopher and social theorist who proposed a positivist philosophy of science. While not exclusively quantitative, his emphasis on empirical observation and measurement laid the groundwork for the later development of quantitative research in the social sciences (Lenzer 1997).

These figures played pivotal roles in advocating for and advancing the use of quantitative methods in their respective disciplines during the eighteenth and nineteenth centuries. Their work laid the groundwork for the increasing integration of quantitative approaches into scientific research and inquiry.

4.3 Qualitative Research Methods

The colonial expansion of the nineteenth century was one of the main reasons to understand the nature of interaction between different cultures, people, and societies. Addressing these issues demanded that different fields interact and collaborate; fields such as natural sciences, archaeology, and social philosophy.

Anthropology as a formal discipline emerged in the nineteenth century as a response to address the issues caused by the expansion of the colonial system. Through systematic observation, study, and analysis, anthropologists document the diversity of human cultures and societies, both past and present. Existing mathematics and quantitative approaches simply did not offer any help in this new science of anthropology, and the problems it addressed. New research methods were necessary to understand possible links between cultural and social features in these newly discovered societies. In the end, necessity forced qualitative methods to emerge, but much later than quantitative methods, and as a response to the limitations of purely quantitative approaches (Malinowski 1922, 1989; Mead 1949).

Qualitative research methods, such as participant observation and case studies, have roots in the social sciences. Anthropologists like Bronisław Malinowski

(Kluckhohn 1943; Firth 2023), Margaret Mead, and sociologists associated with the Chicago School (late nineteenth to early twentieth centuries) played a crucial role in developing qualitative methods.

Bronislaw Malinowski (1884–1942) was a pioneering anthropologist whose work significantly contributed to developing qualitative approaches in social sciences, particularly in anthropology. Malinowski is often regarded as one of the founders of modern ethnographic research, which involves the immersive study of cultures through direct interaction with the subjects being studied. In his book (Malinowski 1922), Malinowski is credited with popularizing the method of **Participant Observation** in ethnographic research. He emphasized the importance of firsthand experience and direct engagement with the community being studied. Malinowski is known for traveling to Australia to understand Aboriginals' mentality, lifestyle, and cultural custom, believing that living among the people being studied was essential for gaining a deep and holistic understanding of their culture, customs, and daily lives.

Malinowski introduced the **functionalist** perspective to anthropology, which focused on understanding how various cultural elements and practices served specific functions within a society. The functionalist perspective was presented in a book (Malinowski 1989) where he emphasized the idea that cultural practices and institutions exist to fulfill basic human needs and contribute to the stability and functioning of a society. This book explores the relationship between social norms and crime in indigenous societies, offering insights into Malinowski's perspective on the functional role of cultural practices. Malinowski advocated for a holistic approach to studying cultures. He believed that every aspect of society, from economic practices to religious rituals, should be studied in relation to one another to indicate the interconnectedness and interdependence of different cultural elements.

Malinowski emphasized the importance of understanding the emic perspective or the viewpoint of the subjects being studied. He believed that researchers should strive to see the world through the eyes of the people they are analyzing rather than imposing their own preconceived notions or biases. To Malinowski's credit also goes the concept of **in-depth interviews** with informants to gather detailed information about their beliefs, practices, and experiences. This approach allowed him to capture the subjective perspectives of individuals within the culture (Malinowski 1922).

Bronislaw Malinowski's contributions laid the foundation for contemporary qualitative research methods in anthropology and other social sciences. His emphasis on immersive fieldwork, holistic analysis, and understanding the perspectives of the studied community has profoundly influenced how researchers approach qualitative studies, ensuring a deeper and more empathetic understanding of diverse cultures and societies (Firth 2023). These few references provide some of his publications to explore Bronislaw Malinowski's influential contributions to qualitative research and anthropology. His books are testimony of his groundbreaking methods, theoretical perspectives, and immersive fieldwork, which have left a lasting impact on the study of cultures and societies.

Sociology is the scientific study of human society and social behavior. Sociologists were and remain interested in exploring and understanding the social dynamics that shape individuals' actions and interactions, foci of interest, which require methods beyond numerical ones.

One of the most famous sociologists that shaped the future of this science is Max Weber. Max Weber (1864–1920) was a German sociologist, philosopher, and fundamental personality in the development of sociology. Weber played an influential role in shaping qualitative research methods through his ideas, concepts, and approaches. Whereas Weber is often associated with his influential work on sociological theory and methodology, his contributions have had a lasting impact on qualitative research in general.

Some of the most relevant contributions of Weber are presented in the following. Weber introduced the concept of "ideal types," abstract models or typologies used to analyze and compare real-world phenomena. Researchers can use ideal types to categorize and analyze qualitative data, identifying patterns, variations, and underlying structures within a cultural or social context. Weber's concept of "Verstehen," which translates to "understanding" or "interpretation," highlighted the significance of apprehending social phenomena from the perspective of the actors concerned. This idea laid the groundwork for an interpretive approach to research, where the researcher aims to understand the meanings, motivations, and intentions behind human actions and behaviors (Weber 2003). Weber's emphasis on subjective understanding encouraged researchers to delve into individuals' and groups' subjective experiences and motivations. This approach aligns with the focus of qualitative research on the lived experiences and perspectives of the research participants.

Weber's contribution laid the groundwork for the case study method (Weber 2017). This qualitative research approach involves an in-depth examination of a single case or a few cases to understand complex phenomena comprehensively. Case studies allow researchers to explore rich contextual details and analyze the interplay of various factors. *Case study research* is a qualitative research method which involves an in-depth examination of a specific individual, group, organization, or situation, and by so doing understand the complexities and unique characteristics of the subject being studied. Researchers use case studies to investigate complex, real-life phenomena and generate rich, context-specific wisdom. Weber's work laid the groundwork for qualitative research methods like case studies. Today the term "case study method" has been more fully developed and articulated by later scholars in the field of social sciences.

Margaret Mead (1901–1978) was an American pioneering anthropologist whose work significantly influenced the development and popularization of qualitative research methods. Together with Malinowski, Mead played a crucial role in shaping the field of anthropology and promoting the use of qualitative approaches to study human behavior, cultures, and societies.

Mead is often associated with **participant observation**, which involves immersing oneself in a particular cultural or social setting to gain an in-depth understanding of the people, behaviors, and dynamics under the study. Mead conducted extensive fieldwork in Samoa and other Pacific islands, living among the communities she studied (Mead 1928). Her immersive approach emphasized the importance of first-hand experience and direct engagement with research subjects.

Mead is known for introducing the principle of **cultural relativism**. Her work emphasized the idea that different cultures should be understood and interpreted within their own unique contexts rather than being judged by external standards. It would be a mistake to use one culture to judge another culture. Furthermore, it is relevant to point out that different cultures produce different social products. Especially at the time when Margaret Mead lived (1901–1978) and studied social experiments on the isolated island of Australia, this approach was fundamental to understanding the social issues of societies living on those remote islands (Mead 1949). Mead's cultural relativism became a cornerstone of qualitative research, encouraging researchers to suspend their own biases and immerse themselves in the worldviews of the people they were studying.

In the mid-twentieth century, qualitative research began to be more formally recognized as a distinct approach to research. Researchers like Anselm Strauss and Barney Glaser developed grounded theory, and scholars like Howard Becker and Erving Goffman further expanded qualitative methodologies.

Anselm Strauss (1916–1996) was a prominent American sociologist who made significant contributions to the field of qualitative research, particularly in the development of grounded theory. His collaborative work with Barney G. Glaser led to the creation of grounded theory methodology, which has since become a widely used approach in social research.

Strauss and Glaser introduced grounded theory as a research methodology in the 1960s. Grounded theory aims to generate theories grounded in empirical data rather than relying on preexisting theories. It involves systematic data collection, coding, and constant comparison to develop concepts and theories emerging emerge from the data themselves (Chun Tie et al. 2019). Strauss played a crucial role in refining and developing the constant comparative method within grounded theory. This iterative process involves comparing new data with previously collected data to identify patterns, themes, and categories. It helps researchers to refine their concepts and theories based on ongoing data analysis (Hallberg 2006). Glaser and Strauss defended the quality of qualitative research, arguing that grounded theory increased its quality by providing a basis for theory construction (Charmaz and Thornberg 2021).

Prominent sociologist Erving Goffman's contributions to the disciplines of social psychology, communication studies, and sociology have had a long-lasting influence (Goffman 1956). The idea of social contact and how people show themselves in daily life are two of his most significant contributions (Liu 2024). Understanding the subtleties of social behavior and the intricacies of human connection is still largely based on Goffman's work. His observations on how individuals navigate and create social reality are still applicable to sociological studies today (Collins 1980).

Howard Becker's influence on the development and promotion of qualitative research methods in sociology is profound. His unique emphasis on ethnography and the detailed study of social processes from the participants' perspective sets his work apart. This approach provides a deeper understanding of social phenomena, making it a valuable resource for students and scholars in sociology and qualitative research (Segre 2018).

Howard S. Becker's 1963 seminal book "Outsiders: Studies in the Sociology of Deviance" explores the concept of social deviance and how society labels individuals and groups as deviant. Becker discusses the social processes that contribute to these labels and the implications for individuals who are marginalized or considered outsiders (Becker 1963).

Another important group of scientists, which played a pivotal role in developing qualitative research methods, is the Chicago School of Sociology. Scholars like Robert E. Park and William I. Thomas conducted field research in urban environments, examining city dwellers' social dynamics and interactions (Ross 1921; Thomas 1931).

Symbolic Interactionism is a sociological perspective and theoretical framework that focuses on the role of symbols and interactions in shaping human behavior and society. It emerged as a significant theoretical approach within sociology, particularly in the early twentieth century. Whereas it has no inventor, several key sociologists contributed to its development. George Herbert Mead (Zalta et al. 2023) and Charles Horton Cooley (Britannica 2023) are often considered the primary founders of Symbolic Interactionism. Mead and Cooley were American sociologists who explored how individuals create and interpret symbols in their interactions with others, and how these symbols shape their understanding of themselves and of the social world in which they act (Redmond 2015).They represent a significant shift in sociological thinking, emphasizing the importance of understanding human behavior through the lens of social interactions and the meanings individuals attach to these actions (Gerhardt 2000). This approach helped to develop further the qualitative research methods.

Qualitative research methods have a focus on understanding individuals' or groups' subjective experiences, meanings, and interpretations rather than quantifiable data. This approach helped to develop further the qualitative research methods. By embracing these qualitative research methods, anthropology and sociology moved away from strict positivist and quantitative approaches to capture better the richness and complexities of human behavior and social phenomena. These roots have since laid the groundwork for the diverse and thriving field of qualitative research across various disciplines.

4.4 Mixed Research Methods

Historically, a purist view of the differences between quantitative and qualitative approaches (positivist versus non-positivist methods) was the standard for quite some time. After carefully examining the problem under the study, researchers would be challenged to choose one of the two existing approaches: qualitative or quantitative. Mixed-research methods are used when a well-defined and pre-specified research design articulates qualitative and quantitative components while investigating a single research question (Timans et al. 2019). This approach uses the strengths of both qualitative and quantitative approaches to provide a more comprehensive and nuanced understanding of research questions (Regnault et al. 2018; Kaur 2016).

Today mixed methods research has evolved into a third methodological direction. As an approach, it has much to offer health and social science research. Its emergence was a response to the limitations of using one of the two traditional methods: quantitative or qualitative. With time, this third approach became more solid and valuable and by many researchers is now accepted as a legitimate third alternative to these two traditions.

Mixed-methods research has been well-known in the social sciences since the twentieth century and has become popular among professionals of other disciplines (Creswell and Plano Clark 2017; Robinson 2007). The philosophical reinforcement of pragmatism allows and guides mixed methods researchers to use a variety of approaches to answer research questions that cannot be addressed utilizing a singular method (Doyle et al. 2009). Researchers in other disciplines may benefit from such a dynamic approach to address the complex and multifaceted research problems often encountered in the healthcare sector (Halcomb and Hickman 2015).

A good source of information on how to design and use mixed methods research is provided by J. W. Creswell and Plano Clark (2017). This book includes seven mixed methods designs with accompanying journal articles providing details for each design. The authors help readers through the entire research process and present updated examples from published and tested mixed methods studies drawn from various disciplines. This book is in its third edition, showing the great interest of researchers. In addition, the new edition includes information about the dynamic and evolving nature of mixed methods research, four additional methodological approaches, and coverage of new directions in mixed methods (J. W. Creswell and Plano Clark 2017).

Mixed methods research offers a combination of qualitative and quantitative methods, allowing researchers to assemble and analyze textual and numerical data. Thus, this integration goes beyond simply collecting separate data sets; it also involves examining how the two data types interact and contribute to the overall understanding of the problem under study. Researchers can gain a complete picture of the research topic using qualitative and quantitative data together. Qualitative methods provide depth and context, uncovering underlying motivations, perceptions, and meanings, whereas quantitative methods offer breadth and generalizability through statistical analysis (Regnault et al. 2018).

Several authors have contributed to the presentation of fundamental principles and central theoretical tenets of a mixed-methods approach, which involves both quantitative and qualitative data collection in response to research questions (Creswell and Creswell 2017; Dawadi et al. 2021; Halcomb and Hickman 2015; Timans et al. 2019). Other authors have shown the benefits of adopting mixed methods since it integrates post-positivism as well as interpretivism frameworks (Wasti et al. 2022; Dawadi et al. 2021). A critical concept in mixed methods research is known as "triangulation," which refers to the particularity of using different sources, methods, or perspectives to validate, corroborate, and summarize research findings. Triangulation is critical for improving the overall credibility and reliability of the study results (Turner et al. 2017).

One should realize that mixed-methods research is more than just operating two different methods in the same study. Researchers must consider and master their solid underlying research philosophies (also called paradigms). Earlier in this chapter, it is stated that positivism usually underpins quantitative research. Thus, such research is typically based on the assumption that there is one single real world out there that can be measured. The researcher is outside this world and is trying to measure and analyze to make sense of this world. For example, quantitative research would address questions of the type "What proportion of the population in the USA would still vote for Trump in the 2024 elections?"

On the other hand, qualitative research is more likely to be based on interpretivism, meaning it will be based on interviews and focus groups, research which is typically based on the assumption that we all experience the world differently. Since we all have different perceptions of the same world we live in, qualitative research will focus on analyzing the people's interpretations, part of the participants' set. In this case, the qualitative research will ask questions of the type "What does it mean for people to vote?" (Dawadi et al. 2021).

There are at least four significant advantages enjoyed by mixed methods research. The first is the ability to generate a more holistic understanding of the problem under investigation. The second is the ability to enhance validity through the process of triangulation. The third is the increased generalizability. The fourth is the potential to explore unexpected relationships or patterns (Turner et al. 2017).

Mixed methods research also presents challenges. Researchers must have expertise in qualitative and quantitative methods. The complexity of integrating data and findings is another issue researchers must overcome (Timans et al. 2019). Moreover, researchers may need to address the potential for increased resource requirements.

Overall, mixed methods research offers a dynamic and comprehensive approach to addressing research questions that may benefit from both qualitative and quantitative perspectives. It allows researchers to capitalize on each method's strengths while overcoming their limitations.

4.5 Conclusions

Quantitative research is a method in which the researcher sees himself/herself as independent and outside the study he/she is trying to address. In the epistemological context, the researcher is a positivist who only explores the nature of reality. A researcher with a positivist orientation considers reality as being out there in the world and needing to be discovered using conventional scientific methodologies (Bassey 1995).

To summarize quantitative research, we should consider the following:

1. Advantage of objectivity. Quantitative research aims for objectivity by using standardized methods and numerical data, allowing for the replication of studies and verification of findings. Large sample sizes and statistical analyses can lead to broader population generalizations.
2. Advantage of precision. Quantitative methods use precise measurements and allow for statistical analysis of relationships between variables. Quantitative data make it easy to compare different groups, conditions, or time periods on the same measure.
3. Advantage of well-accepted statistical analysis. Quantitative research findings are presented using statistical measures, such as means, percentages, correlations, or effect sizes. They allow for generalizing to larger populations and drawing conclusions based on statistical evidence. Quantitative research provides precise and replicable data, facilitates hypothesis testing and theory building, allows for statistical comparisons and generalizability, and can provide evidence for cause–effect relationships.
4. Disadvantage of response bias, the so-called "courtesy bias." The bias is produced by asking difficult questions that interviewees are uncomfortable answering, generally for political or cultural reasons. For example, the integrity of the collected data interviewing of Muslim women by foreign male journalists regarding a sexually related issue in the presence of their husbands will be in question.

To summarize quantitative research, we should consider the following:

1. Advantage of "ecological validity." Qualitative methods give researchers an in-depth understanding of human behavior, emotions, and social processes in their natural context.
2. Advantage of flexibility. Researchers can adapt their approach, allowing for the exploration of unforeseen aspects and new perspectives. Qualitative research can capture the complexity and holistic nature of social phenomena. Qualitative methods often prioritize the opinions and perspectives of participants, allowing them to play a relevant role in shaping the research.
3. Advantage of generating new thinking. Qualitative research is well-suited for generating new theories and hypotheses (EZEKIEL 2015).
4. Disadvantage of the qualitative method is interpretation, which can introduce a certain level of subjectivity and bias into the analysis. Researcher bias can influence the interpretation of data and the selection of themes. Analyzing qualitative data requires skill and experience, and the results can be more challenging to present and validate.
5. Disadvantage of the qualitative method is generalizability, due to the small sample sizes and the unique nature of each context.
6. Disadvantage if the qualitative method is effort. Qualitative data collection and analysis can be time-intensive due to the need to transcribe, code, and analyze textual or visual data.

The qualitative approach explores causality, while the quantitative approach suggests causality. Both methods must be seen as complementary to each other, not as competitive.

References

Adamson, P. (2008). *Al-Kindi. Stanford Encyclopedia of Philosophy.*

Bassey, M. (1995). *Creating Education Through Research: A Global Perspective of Educational Research for the 21st Century*. Kirklington Moor Press.

Becker, H. (1963). *Outsiders: Studies in the Sociology of Deviance*. Free Press.

Boyer, C. B., & Merzbach, U. C. (2011). *History of Mathematics*. Wiley.

Brezina, C. (2005). *Al-Khwarizmi: The Inventor of Algebra (Great Muslim Philosophers and Scientists of the Middle Ages)*. Rosen Central.

Britannica, T. E. of E. (2023). Charles Horton Cooley. In *Encyclopedia Britannica,*. https://www.britannica.com/biography/Charles-Horton-Cooley

Bryman, A. (1984). The Debate about Quantitative and Qualitative Research: A Question of Method or Epistemology? *The British Journal of Sociology*, *35*(1), 75. https://doi.org/10.2307/590553

Charmaz, K., & Thornberg, R. (2021). The pursuit of quality in grounded theory. *Qualitative Research in Psychology*, *18*(3), 305–327. https://doi.org/10.1080/14780887.2020.1780357

Chun Tie, Y., Birks, M., & Francis, K. (2019). Grounded theory research: A design framework for novice researchers. *SAGE Open Medicine*, *7*, 205031211882292. https://doi.org/10.1177/2050312118822927

Collins, R. (1980). Erving Goffman and the Development of Modern Social Theory. In *The View from Goffman* (pp. 170–209). Palgrave Macmillan UK. https://doi.org/10.1007/978-1-349-16268-0_7

Creswell, J. W., & Creswell, J. D. (2017). *Research design: Qualitative, quantitative, and mixed methods approaches*. Sage Publications, Inc.

Creswell, J. W., & Plano Clark, V. L. (2017). *Designing and Conducting Mixed Methods Research.*

Creswell, John W. (1994). *Research design : qualitative & quantitative approaches*. Thousand Oaks, Calif. : Sage Publications.

Cunningham, J. W., & Fitzgerald, J. (1995). Epistemology and reading. Reading Research Quarterly, 31(1), 36–60. https://doi.org/10.1598/RRQ.31.1.3

Dawadi, S., Shrestha, S., & Giri, R. A. (2021). Mixed-Methods Research: A Discussion on its Types, Challenges, and Criticisms. *Journal of Practical Studies in Education*, *2*(2), 25–36. https://doi.org/10.46809/jpse.v2i2.20

Di Liscia, D. A. (2021). Johannes Kepler. In *The Stanford Encyclopedia of Philosophy*. https://plato.stanford.edu/archives/fall2021/entries/kepler/

Doyle, L., Brady, A.-M., & Byrne, G. (2009). An overview of mixed methods research. *Journal of Research in Nursing*, *14*(2), 175–185. https://doi.org/10.1177/1744987108093962

Eknoyan, G. (2007). Adolphe Quetelet (1796 1874) the average man and indices of obesity. *Nephrology Dialysis Transplantation*, *23*(1), 47–51. https://doi.org/10.1093/ndt/gfm517

EZEKIEL, I. T. (2015). *RESEARCH METHODOLOGY*. PAN AFRICAN UNIVERSITY.

Firth, R. W. (2023). Bronisław Malinowski. In *Encylopedia Britanica*. https://www.britannica.com/biography/Bronislaw-Malinowski

Fox, S. (2016). *'What Exactly Is the Crime?' Paula Findlen Talks About the Life of Galileo.*

Gerhardt, U. (2000). Ambivalent Interactionist: Anselm Strauss and the "Schools" of Chicago Sociology. *The American Sociologist*, *31*(4), 34–64.

Gillham, N. W. (2001). *A Life of Sir Francis Galton: From African Exploration to the Birth of Eugenics*. Oxford University Press.

Gillings, R. J. (2003). *Mathematics in the Time of the Pharaohs*. MIT Press.

Goffman, E. (1956). *The Presentation of Self in Everyday Life.*
Halcomb, E., & Hickman, L. (2015). Mixed methods research. *Nursing Standard, 29*(32), 41–47. https://doi.org/10.7748/ns.29.32.41.e8858
Hallberg, L. R.-M. (2006). The "core category" of grounded theory: Making constant comparisons. *International Journal of Qualitative Studies on Health and Well-Being, 1*(3), 141–148. https://doi.org/10.1080/17482620600858399
Heath, T. (1981). *A History of Greek Mathematics.* Dover Publications Inc.
Hetherington, S. (2019). *What is epistemology?*
Imhausen, A., Robson, E., Dauben, J. W., Plofker, K., & Berggren, J. L. (2007). *The Mathematics of Egypt, Mesopotamia, China, India, and Islam: A Sourcebook.*
Kaur, M. (2016). Application of mixed method approach in public health research. *Indian Journal of Community Medicine, 41*(2), 93. https://doi.org/10.4103/0970-0218.173495
Kessener, P. M. (2021). Roman Water Transport: Pressure Lines. *Water, 14*(1), 28. https://doi.org/10.3390/w14010028
Kluckhohn, C. (1943). Bronislaw Malinowski 1884-1942. *The Journal of American Folklore, 56*(221), 208–219.
Langham, A. (2022). *Pierre de Fermat, Judge Turned Mathematician.* Elephant Learning, LLC. https://www.elephantlearning.com/post/pierre-de-fermat-judge-turned-mathematician
Lenzer, G. (1997). *Auguste Comte and Positivism: The Essential.*
Liu, S. (2024). Social spaces: from Georg Simmel to Erving Goffman. *The Journal of Chinese Sociology, 11*(1), 13. https://doi.org/10.1186/s40711-024-00217-9
Malinowski, B. (1922). *Argonauts of the Western Pacific.* George Routledge & Sons, Ltd.
Malinowski, B. (1989). *Crime and Custom in Savage Society.* ROWMAN & LITTLEFIELD.
Mead, M. (1928). *Coming of Age in Samoa.* https://worldhistorycommons.org/margret-mead-coming-age-samoa
Mead, M. (1949). *Male and female: a study of the sexes in a changing world.*
Mosselmans, B. (2015). *William Stanley Jevons and the Cutting Edge of Economics.* Routledge.
Rashed, R. (2010). *Al-Khwarizmi: The Beginnings of Algebra (History of Science and Philosophy in Classical Islam).* Saqi Books; Illustrated edition.
Redmond, M. V. (2015). *Symbolic Interactionism.*
Regnault, A., Willgoss, T., & Barbic, S. (2018). Towards the use of mixed methods inquiry as best practice in health outcomes research. *Journal of Patient-Reported Outcomes, 2*(1), 19. https://doi.org/10.1186/s41687-018-0043-8
Robinson, P. (2007). Designing and Conducting Mixed Methods Research. *Australian and New Zealand Journal of Public Health, 31*(4), 388. https://doi.org/10.1111/j.1753-6405.2007.00096.x
Ross, E. A. (1921). Introduction to the Science of Society. Robert E. Park, Ernest W. Burgess. *American Journal of Sociology, 27*(3), 289–432.
Segre, S. (2018). The Sociology of Howard S. Becker: Theory with a Wide Horizon. *Contemporary Sociology: A Journal of Reviews, 47*(6), 665–667. https://doi.org/10.1177/0094306118805417a
Tent, M. B. W. (2006). *The Prince of Mathematics: Carl Friedrich Gauss.*
Thomas, W. I. (1931). The Relation of Research to The Social Process. *Essays on Research in the Social Sciences*, 175–194.
Timans, R., Wouters, P., & Heilbron, J. (2019). Mixed methods research: what it is and what it could be. *Theory and Society, 48*(2), 193–216. https://doi.org/10.1007/s11186-019-09345-5
Trow, M. (1957). Comment on "Participant Observation and Interviewing: A Comparison". *Human Organization, 16*(3), 33–35. http://www.jstor.org/stable/44127709
Turner, S. F., Cardinal, L. B., & Burton, R. M. (2017). Research Design for Mixed Methods. *Organizational Research Methods, 20*(2), 243–267. https://doi.org/10.1177/1094428115610808
Urvoy, D. (2017). *Ibn Rushd (Averroes).* Routledge, Taylor & Francis.
Viegas, J. (2006). *Al Kindi: Father of Arab Philosophy (Great Muslim Philosophers and Scientists of the Middle Ages).* Rosen Central.

Wang, J. (2020). What is Epistemology? *The Philosophical Quarterly*, *71*(1), 225–226. https://doi.org/10.1093/pq/pqaa032

Wasti, S. P., Simkhada, P., van Teijlingen, E., Sathian, B., & Banerjee, I. (2022). The Growing Importance of Mixed-Methods Research in Health. *Nepal Journal of Epidemiology*, *12*(1), 1175–1178. https://doi.org/10.3126/nje.v12i1.43633

Weber, M. (2003). *The Protestant Ethic and the Spirit of Capitalism*. Dover Publications Inc.

Weber, M. (2017). *The Methodology of the Social Sciences*. Taylor & Francis.

Wolpert, L. (2002). *Science and Mathematics in Ancient Greek Culture* (C. J. Tuplin & T. E. Rihll (eds.)). Oxford University Press. https://doi.org/10.1093/acprof:oso/9780198152484.001.0001

Zalta, E. N., Nodelman, U., Aboulafia, M., & Taylor, S. (2023). George Herbert Mead. In *The Stanford Encyclopedia of Philosophy*. https://plato.stanford.edu/archives/spr2023/entries/mead/

Chapter 5
Introduction to Artificial Intelligence

5.1 Introduction

Artificial intelligence is not a product of mere technological advancement but a result of a deep examination of the cognitive process and the human brain. By identifying patterns in the brain and learning about its functions, AI has made noteworthy progress. These investigations have paved the way for the development of intelligent systems and software.

Cognitive psychology is the branch of psychology that studies how people think, the interaction of thinking, emotion, creativity, and problem-solving abilities affect how and why you think the way you do (Solso et al. 2005). The fascinating and beneficial union of cognitive psychology and technology has established the groundwork for further scientific advancements. The evolution of technology and human cognitive abilities are closely related, as seen by the history of technological advancement. The history of humanity demonstrates the idea that humans shape technologies, which also shape us in return (Osiurak et al. 2018). The authors offer a framework for comprehending how technology both shapes and is shaped by human cognition. They suggest three phases for the interaction between technology and cognitive skills: the past, present, and future.

5.1.1 The Past

The initial stage, known as the Past, was defined by the use of physical tools. Increasing our sensorimotor abilities is what physical tools are all about (Virgo et al. 2017). People still use these tools extensively, even though they are comparable to the earliest tools that humans built and utilized in prehistory. It is important to note that using any physical tool requires the user to comprehend the geometry and

P. Papajorgji, H. Moskowitz, *The Mind of Everyday*,
https://doi.org/10.1007/978-3-031-78078-3_5

physical attributes of the tool on a cognitive level. Features of early stone tools show that their makers were aware of the technique needed to make them (Hovers 2011).

5.1.2 The Present

The application of advanced tools characterizes the second phase, the Present. Today we use sophisticated tools to accomplish everyday tasks. Good examples are sending messages via a mobile device or traveling across the city on a subway. When using complex, it is typical for the user to require some time and perhaps special training to become familiar with the way they operate. With the increasingly sophisticated cognitive tools (like computer spreadsheets) of tools, there is a significant disconnect between creating and using a tool. Even though the tools are more complex, the way people use the tool has not changed. The result turns out to be a change in the meaning of the tool; the tool ends up meaning the user's objective, rather than the tool itself.

5.1.3 The Future

Using symbiotic tools characterizes the third phase, which is the **Future**. What kind of tools there will be in the future is difficult to predict. They are frequently featured in science fiction films. Among those already developed, the business world ends up adopting some whereas others are never implemented. Several upcoming tools fall under the category of brain–computer interfaces (BCIs), which have attracted much attention since their goal is to convert brain activity (or "thoughts") into commands that a machine can comprehend. This process is achieved using an intelligent algorithm; smart sensors record and evaluate brain activity and then correlate it with the appropriate action performed by the artificial system. This algorithm, as demonstrated by Bishop (2007), is capable of learning to distinguish classes in the brain signals recorded. The true power of an intelligent BCI interaction lies in its adaptability. It includes a learning phase, which can adjust the technology to the unique cognitive system of each user. The personal characteristics of our brains and the functional differences among us stop being obstacles but rather encouraging factors, which drive the earning algorithms to be personalized to the characteristics of each individual. This adaptability is a testament to the potential of BCIs to "seemingly understand" and by doing so enhance our individual experiences.

Scholars in philosophy, psychology, and cognitive science were fascinated by the role that humans play in the explanation process. They looked into how people identify, produce, choose, assess, and give explanations. What they found is that social expectations and cognitive biases are major factors in this process (Miller, 2018).

5.1.4 Founding AI

The development of artificial intelligence occurred in the 1940s and 1950s. A conversation regarding the potential for developing an artificial brain was started by scientists from a variety of disciplines, including mathematics, psychology, engineering, economics, and political science. The field of artificial intelligence was established in 1956 as a research area (Kaplan 2022).

In 1955, John McCarthy, at the time, an Assistant Professor of Mathematics at Dartmouth College, Hanover, New Hampshire, USA, organized a group of researchers to clarify the concepts and develop ideas about thinking machines. While trying to baptize this new field with a meaningful name, McCarthy selected "Artificial Intelligence." Early that year, McCarthy had made a request to the Rockefeller Foundation to fund a summer seminar at Dartmouth for about ten participants. The Dartmouth Workshop of 1956 was organized by Marvin Minsky, John McCarthy, and two senior scientists: Claude Shannon, the father of information theory then working at Bell Labs, and Nathan Rochester of IBM (Kline 2011). The text justifying the need for the conference included "every aspect of learning or any other feature of intelligence can be so precisely described that a machine can be made to simulate it" (Crevier 1993).

The conference included many famous researchers such as Ray Solomonoff, Oliver Selfridge, Trenchard More, Arthur Samuel, Allen Newell, and Herbert A. Simon, each of whom made an essential contribution during the first decades of AI research (Crevier 1993). At the workshop, Newell and Simon argued that the name of the new field should be the "Logic Theorist."

Logic Theorist is a computer program written in 1956 by Allen Newell, Herbert A. Simon, and Cliff Shaw. It was the first program deliberately engineered to perform automated reasoning, and has been described as "the first artificial intelligence program" (Crevier 1993). Logic Theorist proved 38 of the first 52 theorems presented in the famous book *Principia Mathematica*, authored by Alfred Whitehead and Bertrand Russell and for some theorems even found new and more elegant proofs (Whitehead and Russell 2004). McCarthy's arguments persuaded the attendees to use "Artificial Intelligence" as the name of the field (McCorduck 2004). The 1956 Dartmouth workshop was the moment that AI gained its name, defined its mission, made an inventory of its first success and its major players, and is widely considered the birth of AI (McCorduck 2004).

Alexander Ivakhnenko made significant contributions to the development of artificial intelligence, particularly in the field of machine learning. He was a prominent Ukrainian mathematician and computer scientist known for his work on the development of the Group Method of Data Handling (GMDH) algorithm.

Ivakhnenko's GMDH algorithm, introduced in the 1960s, was a pioneering technique in the field of machine learning. It automatically constructs mathematical models from empirical data without relying on prior assumptions or explicit model structures. The algorithm's key idea was to iteratively build and refine models based on a training dataset, progressively improving their accuracy and predictive capabilities. The GMDH algorithm implemented a bottom-up approach, starting with a set of input variables and gradually generating higher-order terms and combinations to construct the model. It utilizes a data-driven approach to identify and select the most relevant variables and interactions, adapting the model complexity to the specific dataset. This capability makes GMDH well-suited for tasks such as regression analysis, pattern recognition, and prediction.

Ivakhnenko's work on the GMDH algorithm significantly impacted the development of artificial intelligence and machine learning. His approach provided a flexible and adaptive method for extracting knowledge from data, enabling automated modeling and decision-making. The GMDH algorithm found applications in diverse fields, including engineering, economics, medicine, and more. Ivakhnenko's contributions to artificial intelligence extend beyond the GMDH algorithm. His research explored the potential of neural networks in modeling complex systems and understanding human cognitive processes. He also made advancements in the areas of pattern recognition, neural networks, and cognitive sciences.

Overall, Alexander Ivakhnenko played a crucial role in the development of artificial intelligence by introducing innovative machine-learning techniques, notably the GMDH algorithm. His work paved the way for data-driven modeling and automated knowledge extraction from empirical data, contributing to the advancement of AI and its applications in various domains.

5.2 Reactive Machines

Reactive machines refer to a class of intelligent systems designed to react and respond to specific stimuli or inputs in real time without any internal representation or memory of past events (Brooks 1986; Brooks 1991). These machines operate solely based on the immediate sensory information they receive, enabling them to make rapid and adaptive decisions in dynamic environments.

Unlike traditional computing systems which rely on predefined rules or stored data, reactive machines are built on reactive architectures, which emphasize direct stimulus–response connections. They use sensors and actuators to perceive the state of their environment, and in turn produce appropriate outputs in real time. This immediate feedback loop enables reactive machines to exhibit responsive and interactive behavior without the need for explicit planning or internal models of the world.

Reactive machines often incorporate techniques from the field of artificial intelligence, including machine learning and pattern recognition (Ivakhnenko and Ivakhnenko 2009), enabling them to learn and adapt their behavior based on experience. Reactive machines can be trained using various algorithms such as reinforcement learning, where they learn to associate certain actions with positive or negative outcomes. This enables them to optimize their responses and make decisions leading to favorable outcomes in each context. One of the notable characteristics of reactive machines is their ability to handle uncertainty and ambiguity in real-time situations. They can dynamically adjust their behavior based on the changing conditions of their environment, allowing them to cope with unpredictable events and operate in complex and dynamic scenarios (Arkin 1998; Nilsson 1994).

Reactive machines have found applications in various domains, including robotics, autonomous vehicles, and real-time decision-making systems. They can be employed in scenarios where real-time responsiveness and adaptability are critical, such as industrial automation, autonomous navigation, and interactive virtual agents (Vernon et al. 2007; Mondada et al. 2010). They can be used successfully in the education system (Araújo et al. 2015).

To summarize, reactive machines are intelligent systems which react and act upon immediate stimuli without the need for explicit internal models or prior knowledge. Their ability to make quick, flexible judgments based on sensory input and real-time feedback loops enables them to function well in unpredictable and changing settings. The Deep Blue and Netflix Systems are two instances of reactive machines to be discussed in the following sections.

5.2.1 Deep Blue

The efforts to design and develop computers able to play and beat chess champions started in the 1980s at Carnegie Mellon University, USA. Two machines were designed at the time, ChipTest and Deep Thought. Both systems used a single-chip chess move generator (Hsu 1987) to achieve search speeds in the vicinity of 500,000 positions per second (ChipTest) to 700,000 positions per second (Deep Thought).

In January of 1988, at a press conference in Paris, a journalist asked world chess champion Gary Kasparov, whether a computer would have a chance to beat a chess grandmaster before 2000. Kasparov's answer was firm. "No way," he said, adding, "If a grandmaster has difficulties playing with a computer, I would be happy to provide my advice." Time was not kind to Kasparov. Only 10 months later, a computer designed in a graduate project at Carnegie-Mellon University referred to as Deep Thought, beat the grandmaster Bent Larsen in a public show. Bent Larsen was a contender for the world title at the time (Hsu et al. 1990).

Kasparov made headlines worldwide when he lost the second of two chess matches against the IBM supercomputer Deep Blue. In the wake of that loss, he began thinking and writing about the future of artificial intelligence, arguing that the future of AI was not so much about machines replacing humans in the workforce as it was about intelligent machines learning to collaborate with humans (Kasparov 2022).

Deep Blue was developed by a team of researchers and engineers at IBM, led by Feng-hsiung Hsu. The project started in the mid-1980s and aimed to build a computer capable of challenging and defeating human chess players. Deep Blue utilized a custom-designed parallel processing architecture consisting of multiple high-speed processors. Deep Blue's hardware was specifically optimized for evaluating and analyzing chess positions rapidly.

The software running on Deep Blue employed sophisticated algorithms and heuristics to assess chess positions, calculate possible moves, and determine the best course of action. The program utilized extensive chess databases and employed deep search techniques to explore potential moves and counter-moves (Shanon 1950).

Deep Blue's most famous encounter occurred in 1997 when it faced Garry Kasparov, the reigning world chess champion, in a rematch. The previous year, Deep Blue had lost the match to Kasparov. However, in the rematch, Deep Blue emerged victorious, winning two games out of six and drawing the other four. This made Deep Blue the first computer to defeat a reigning world chess champion in a match.

Deep Blue's victory over Kasparov was a major milestone in artificial intelligence and computer chess, demonstrating that computers had reached a level of skill and strategic thinking that could surpass human champions in certain domains. The match also sparked debates about the capabilities of machines and their potential impact on human intelligence. Deep Blue's success paved the way for further advancements in computer chess and artificial intelligence. It inspired researchers to develop more sophisticated chess engines and led to the creation of more powerful chess-playing programs. Deep Blue's technology and techniques also influenced the development of other AI applications beyond chess.

Overall, Deep Blue was a groundbreaking computer, which made history by defeating Garry Kasparov and showcasing the growing capabilities of AI in strategic games like chess. It remains an iconic milestone in the field of artificial intelligence and continues to inspire advancements in machine learning and computational intelligence (Murray et al. 2002).

There was a vivid discussion later after Deep Blue's victory versus Kasparov on the main reasons Deep Blue performed so magnificently against the world chess champion. According to (Murray et al. 2002) the reasons are:

- A single-chip chess search engine.
- A massively parallel system with multiple levels of parallelism.
- A strong emphasis on search extensions.
- A complex evaluation function, and.
- Effective use of a Grandmaster game database.

A few people were wondering why investing so much money and time to design and implement a machine that would play chess is so relevant. Claude Shannon, the father of Information theory, gave the simple but profound answer, "The investigation of the chess playing problem is intended to develop techniques that can be used for more practical applications" (Shanon 1950).

5.2.2 *Recommender System*

The Internet, be it for personal or work-related reasons, is a necessity that nowadays cannot be missed. One can work and have a decent standard of living only with access to the Internet. Anytime one uses the Google Search system or any other search engine, the system "remembers" what it is searching for and systematically collects and stores this information for future uses. As early as the end of the last century, efforts to introduce recommender systems in e-commerce were taking place (Schafer et al. 1999). The number of digital information available on the Internet and the number of users using the Internet to search for information has dramatically increased to create information overload challenges (Isinkaye et al. 2015).

One solution was designing robust storage systems that partially solved the problem. It became quickly obvious that on should develop a system, which considered the context and could store users' preferences to use them for future searches. Such systems are called recommender systems. Recommender systems are designed to filter the collected information helping to deal with the problem of information overload. The filtering process deals with a large amount of dynamically generated information according to user preferences or interests. When making its prediction, the recommender system takes into account the user's profile.

5.3 ChatGPT

In November 2022, the world was effectively "turned upside down" by the emergence of a what can be rightly described as a unique, extremely important technological development, the ChatGPT product of OpenAI company. The acronym "CHATGPT"refers to "Conversational Hypertext Access Technology Generative Pre-trained Transformer." This term is not a standard or widely recognized acronym but is a way to describe the underlying technology and purpose of models like ChatGPT (OpenAI 2023).

Natural languages are difficult, not précised languages. They are not mathematics and difficult to find mathematical relationships among elements of a sentence. The words may appear in different places within the sentence and at times, their position may alter the meaning of the sentence. One must consider the entire sentence before capitalizing in the meaning of it. In other words, the sentence's meaning is context-bound.

ChatGPT can be best understood as the latest instantiation of efforts to understand thought, efforts which appreciated the complexity of biology, and the need to simulate biology using mathematics. The story begins with the efforts to discover mathematical relationships among elements of the sentence that would lead to a computer-based system to understand text. These efforts were not successful, despite their mathematical elegance.

Researchers then turned toward understanding how the human brain works an effort with a long history. The history begins with the nineteenth century efforts to understand how the human brain works. At this time, the majority of scientists thought the brain was made up of an endless web of fibers as twisted as a labyrinth. Santiago Ramón y Cajal presented the first unequivocal evidence that the brain is constituted of individual cells, subsequently named neurons, that are basically the same as those that make up the rest of the living world (Ehrlich 2022).

It is commonly believed that Cajal founded contemporary neuroscience. Historians have ranked him alongside Darwin and Pasteur as one of the greatest biologists of the nineteenth century and, among other famous scientists such as Copernicus, Galileo, and Newton, one of the most outstanding scientists of all time. His masterwork, Texture of the Nervous System of Man and the Vertebrates, is a foundational neuroscience publication comparable to Darwin's **On the Origin of Species. F**or evolutionary biology. Cajal was awarded the Nobel Prize in 1906 for his work on the structure of neurons (Ehrlich 2022).

Among steps to understand how the human brain functions stands the work of the 2014 Laureates, which radically altered our understanding of these functions. John O'Keefe discovered place cells in the hippocampus that signal position and provide the brain with spatial memory capacity. May-Britt Moser and Edvard I. Moser discovered that in the medial entorhinal cortex, a region of the brain next to the hippocampus, grid cells provide the brain with an internal coordinate system essential for navigation. This discovery states that different parts of the human brain are specialized in performing different functions (Institutet 2014).

5.4 Neural Networks

To understand how CHATGPT works, we now turn to Neural Networks. The human brain and the network of neurons present in the brain inspired the birth and development of a new science referred to as Artificial Neural Networks (ANNs).

Artificial Neural Networks mimic the network of neurons in the mammalian cortex. Whereas ANNs are based on the mammalian system, they are really modeled on a much smaller scale. The mammalian brain system contains billions of neurons, whereas in ANNs, there are hundreds or thousands of processor units, a tiny number compared to the mammalian brain structure (Nwankpa et al. 2018). ANNs can be implemented as algorithms or as an actual piece of hardware.

According to the inventor of one of the first neurocomputers, a neural network can be defined as: ...***a computing system comprising several simple, highly interconnected processing elements that process information by their dynamic state response to external inputs*** (Siddharth et al. 2020).

Currently, researchers model ANNs as Input Layer, Hidden Layer, and Output Layer, even though a neural network typically has more than one hidden layer (see Fig. 5.1.)

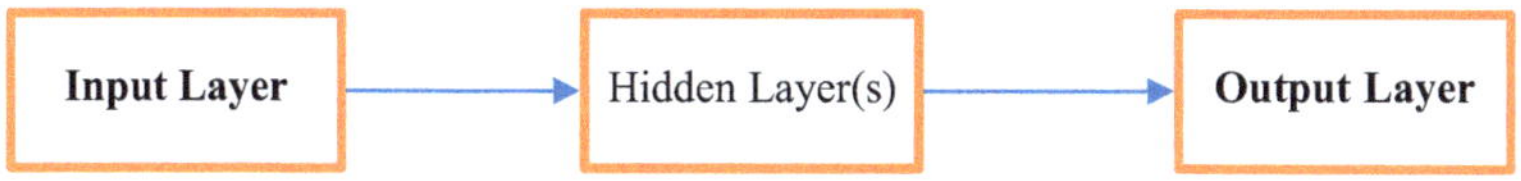

Fig. 5.1 General structure of ANNs

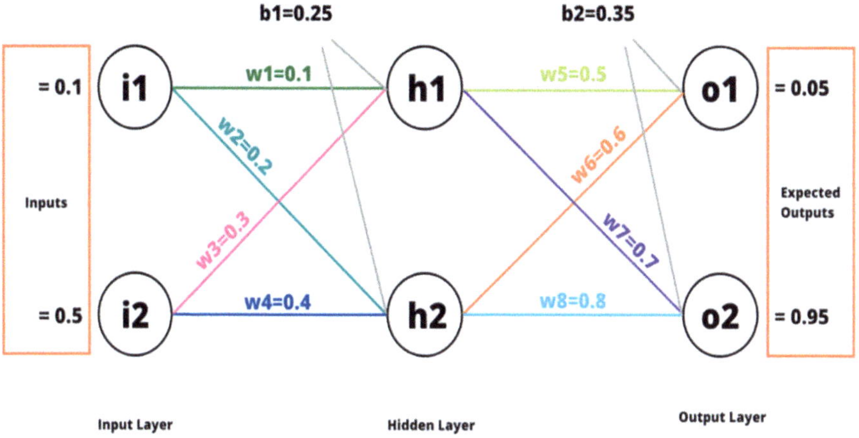

Fig. 5.2 Example of a neural network. This image is provided by Lamsal (2021)

The information is processed and passed on from one neuron to another in the human brain through neurosynaptic junctions. Similarly, in artificial neural networks, there are different layers of cells arranged and connected, as shown below in Fig. 5.2.

The output/information from the inner layers of the neural network becomes inputs to the following layers. This process continues until it finally reaches the outermost layer, which delivers the final output. The input to the outer layer provides nonlinearity to the inner layers' output so that it can be further processed. A function performs the output of a unit, referred to as the activation function.

Activation functions play a pivotal role in neural networks. Without them, the output signal would be a simple linear function, just a polynomial of degree one. Whereas a linear equation is straightforward to solve, its complexity is limited, and it cannot learn and recognize complex mappings from data. This limitation is why a neural network without activation functions as a Linear Regression Model with limited performance and power in most cases (Emanuel et al. 2023). It is clear that activation functions are crucial for neural networks to address complex tasks such as modeling complicated data types like text, images, videos, audio, and speech (Ramachandran et al. 2017).

Nonlinearity refers to the nonlinear function, often called an activation function, that is applied to the linear output of a neuron. This function introduces nonlinear properties to the model. Nonlinearity is a crucial feature of neural networks. Without nonlinear activation functions, regardless of the number of layers a neural network is composed of, the entire network would behave like a single-layer linear model. The composition of linear functions is still linear. In the real world, many relationships are nonlinear. With nonlinear activation functions, neural networks can capture and learn these patterns. Thus, nonlinearities allow neural networks to approximate complex, nonlinear functions.

Deep neural networks with nonlinearities can learn features with a hierarchical structure. In the area of image recognition, lower layers might learn to detect edges, the following layers might learn to detect shapes by combining edges, and even higher layers might detect more complex structures. This hierarchy of feature extraction is possible due to the nonlinear transformations.

A field where the nonlinearity feature is used with great success is Function Approximation. Neural networks can serve as universal function approximators, and the use of nonlinearity is a crucial factor in allowing this capability (Taherdoost 2023). Another meaningful use of the nonlinearity feature is in classification. On-linearities allow networks to draw complex decision boundaries, making them appropriate for a variety of classification tasks, including binary as well as multi-class problems. In cases where data have nonlinear relationships between inputs and outputs, neural networks provided with nonlinear activations may perform better than traditional linear methods (Heil et al. 2023).

Advanced neural network models like Convolutional Neural Networks (CNNs) used in Deep Learning Architectures for image recognition or Recurrent Neural Networks (RNNs) for sequence data rely on nonlinear features to process and generate hierarchical data representations (Nanni et al. 2021). The nonlinearity is crucial to designing and implementing neural networks because it increases the expressive power and utility of deep learning. Nonlinearity ensures that the network can learn from errors and make adjustments, essential for learning complex patterns (Emanuel et al. 2023).

The most appealing property of Artificial Neural Networks is the ability to adapt their behavior according to the changing characteristics of the system. In the last few decades, researchers and scientists have investigated methods to improve the performance of Artificial Neural Networks. Such as optimizing the training methods, hyperparameter tuning, learning parameters, or network structures (Siddharth et al. 2020).

5.5 Self-Driving Car

Let us illustrate how neural networks are used to drive a car. The car has distance sensors located on the four sides, front, back, left and right, respectively. Proximity sensors continually provide distance values on all sides of the car. The proximity sensors report a value of 1.0 when something is very close and 0.0 when no vehicle is detectable nearby.

The simplified diagram (Fig. 5.3) shows the major components of an autonomous vehicle's neural network system and how they interact with each other. The figure shows the arrangement, and how the inputs interact with each other.

1. The Input Component containing various sensors (like cameras, LIDAR, and radar) gathers real-time data about the vehicle's surroundings, such as cars, gates, and other traffic obstacles.
2. The Data Preprocessing Component is where the raw data from sensors is stored, processed, and converted into a format suitable for the neural network.
3. The Neural Network component is the core element, which processes the input data. The Neural Network usually comprises several layers, each layer designed to identify and interpret different aspects of the data. Example of these aspects are identifying objects, estimating distances, and predicting the movement of other surrounding entities.
4. Neural networks make decisions based on the processed data, such as steering, accelerating, or braking. Thus, there is a Decision-Making component covering all these issues.
5. As explained above, the decision-making component decides what the system should do. However, there is a need for a controlling component, which physically conveys the decision to the steering control, brakes, etc. Thus, the Control component is a necessary part of the system.
6. Driving a car means collecting, analyzing, and making decisions constantly. Thus, a Feedback Loop component is necessary to adjust the decision-making process based on the data collected and analyzed by the system.

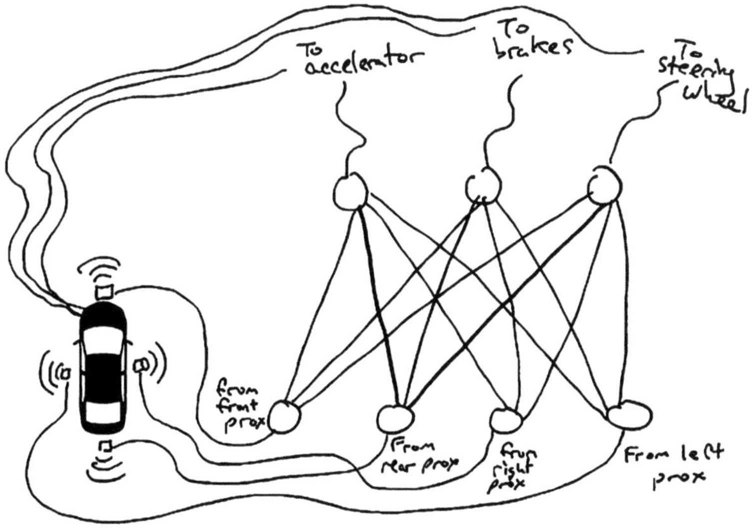

Fig. 5.3 A simplified neural network diagram for a self-driving car. (Riedl 2023)

In order to clarify the ideas of how a neural network-based system self-driving car works, a simplified example is shown in Fig. 5.3. An accurate diagram presenting a self-driving car is more complex and sophisticated, but it would be cumbersome to present such a diagram.

Figure 5.3 shows that the car has a distance sensor in the front, back, and both sides, left and right. Distance sensor data are the input for the neural network. In turn, the output from the neural network is connected to actuators, which control and command the brake system, the acceleration, and the steering wheel system.

The car has a robotic mechanism or actuators. Actuators are devices used to convert energy into motion. They are essential components in a large number of systems where mechanical action is needed. Actuators are robots "arms and muscles," providing movement for tasks. The energy used by an actuator can be electrical, hydraulic, pneumatic, or even human power. The motion generated can be linear or rotational. In vehicle systems, actuators perform braking, throttle control, and other automotive systems.

When the accelerator receives a value of 1.0, it uses maximum acceleration. In turn when the accelerator receives no information, viz., a value of 0.0, nothing is done. Similarly, a value of 1.0 sent to the braking mechanism will mean to brake hard, and 0.0 means no braking. The steering mechanism takes a value of −1.0 to +1.0, with a negative value meaning steer left while a positive value means steer right and 0.0 means no turning, keep driving straight.

The system needs to have a recorded "logic of driving," requiring us to convey to the system what we do when we drive. For example, when there is no car in front of us, and the car's speed is lower than the allowed speed the car could accelerate. When a car is front of us is within a critical distance, our car needs to use the brakes and slow down. When a car following us gets too close, and there is no car ahead of us, the car needs to accelerate. Thus, our driverless car needs to learn many driving configurations based on the data sensors send to the Data Preprocessing Component.

Now, the system comprising the car, distance sensors, and actuators is ready to get a taste of the road. At some point, other vehicles of all types will surround our car system. Distance sensors will provide data from all sides of the car, electric current will flow from sensors to all actuators, and the car brakes, accelerates, and turns left and right simultaneously. The result is inevitable disaster, and the disaster is inevitable. Figure 5.4 shows a system where all actuators are activated at the same time.

Thus, it is clear that not all actuators should operate simultaneously. Our system should have some special devices on different parts of the circuits to control electricity flowing more freely between specific sensors and particular actuators. In the scenario that another vehicle is in front at a critical distance and at the left and right sides are other vehicles, it is necessary to have an electric signal only to the brake actuator that will cause our car to slow down and no signal to the steering system. Thus, some actuators have to receive electric signals, and others do not. Some other actuators must be put on a waiting status until the signal reaches a particular value. This system needs a device named ***gate*** used to accumulate the electric signal up to a specific value and only then let it go to the next level.

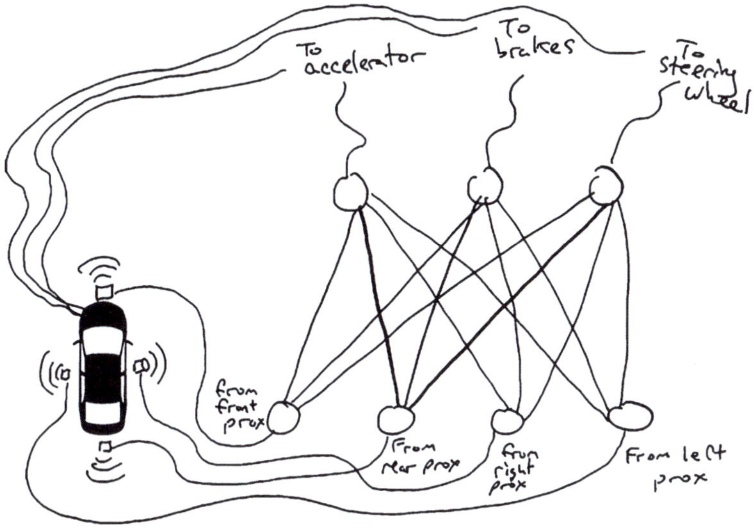

Fig. 5.4 All actuators are activated simultaneously. (Riedl 2023)

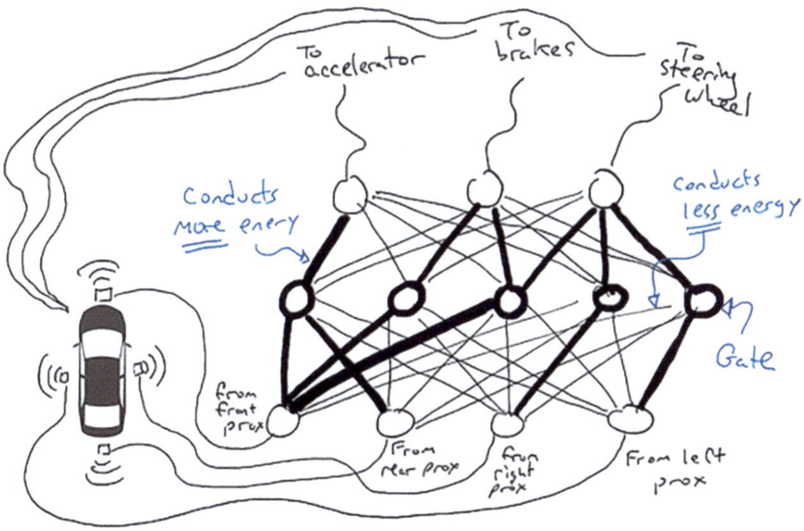

Fig. 5.5 The neural network driving the car. (Riedl 2023)

Figure 5.5 shows a complete neural network that can be trained to send the right kind of message at the right time to the right actuator. Training means simulating most of the road scenarios and the right solution to avoid the disaster.

5.6 Large Language Models

A large language model (LLM) like GPT-4, GPT-3, BARD, Claude, etc., is an advanced type of artificial intelligence system designed to process, understand, and generate human language that is coherent, contextually relevant, and often not very different from human-written text.

LLMs are based on deep learning and neural network architectures, primarily transformer models, which allow them to process and produce text in a way often contextually appropriate and syntactically legible. For example, the autocomplete feature, or the ability to predict the next word in a sentence, is provided by a language model (Zhao et al. 2023; Blank 2023; Naveed et al. 2023; Linegar et al. 2023; Meyer et al. 2023).

The goal of LLM's is to find the output word, which follow the given input words. For example, suppose we have entered the words "The king sits on the ______" and would like to guess what word will replace the space. When someone reads the sentence, they immediately think of the word "throne" and not "sofa." In other words, the word "sofa" has almost a probability of 0, and "throne" has a probability of 1 to be used, knowing that we have already entered the words "The," "king," "sits," "on," "the." Large language models use a probabilistic approach to guess the most appropriate output word, knowing some input words. Thus, in this case, we will write:

$$\boldsymbol{P(throne \mid The, king, sits, on, the)}$$

The system will calculate the probability of having the word "throne" knowing already the words "The," "king," "sits," "on," and "the." If the system delivers a higher probability to the word "sofa" than to "throne," then it is not a sound system; "throne" should receive a much higher probability than "sofa."

This argument can be generalized as follows:

$$P\left(word_n |, word_1 |, word_2 |, \ldots |, word_{n-1}\right)$$

means calculate the probability of the n-th word in a sequence given all the words which come before it (words in positions 1 through $n - 1$).

5.6.1 *Transformers*

The concept of the transformer model was introduced in 2017 by a group of scientists at Google Brain, a deep learning artificial intelligence research team at Google, focusing on deep learning research and applications. Google Brain is a significant player in promoting the field of AI and has been at the vanguard of developing several groundbreaking technologies and models, particularly in deep learning and natural language processing. The development of the Transformer model is a remarkable illustration of their work.

Transformer is a type of neural network architecture that has gained significant attention since its introduction in the paper "Attention Is All You Need" (Vaswani et al. 2017). It has become enormously significant in the field of natural language processing (NLP) but is also being adapted for use in other areas (Khan et al. 2023), (Choi and Lee 2023). The most relevant innovation of the Transformer model is the use of an attention mechanism.

Before the birth of the transformer, some neural networks such as RNNs (Recurrent Neural Networks) or LSTMs (Long Short-Term Memory Networks) were used to analyze and process the input data sequentially. Because of the sequential nature of the treatment, they had quite some limitations while processing data and, thus, had difficulties understanding long-range dependencies in the data. The attention mechanism in the Transformer (Vaswani et al. 2017) enables the model to focus on different parts of the input sequence when generating each part of the output sequence.

To illustrate how a Transformer works, let us consider an example of machine translation. The task is to translate a sentence from English to Italian using a Transformer model. The Transformer model's ability to analyze and handle long-range dependencies and nuances in language makes it particularly effective for tasks like translation. We are simplifying the description of transformers to make it easy for our social sciences audience.

Translating **"Today is a beautiful day**" from English to Italian using a Transformer model involves several operations. Let us break down these operations step by step, remembering that the actual computational processes are complex and apply elaborate mathematical activities, which need not concern us here.

The first step is Preprocessing. The input sentence: "Today is a beautiful day" goes through the tokenization process, meaning divide the input sentence tokens: ["Today," "is," "a," "beautiful," "day"].

The second step is Embedding and Positional EncodingWord Embeddings (Chen et al. 2021; Dufter et al. 2022). Each token is transformed into a high-dimensional vector, which represents its semantic and syntactic features. Semantic features capture the meaning of a word. For instance, the word "president" is associated with government, power, and leadership meanings. Semantics also apply abstract concepts and conceptual information, such as emotions or abstract ideas conveyed by words (Kane et al. 2022; Tabinda Kokab et al. 2022). In contrast, Syntactic Features include information about whether a word is a noun, verb, adjective, etc. For instance, "write" is a verb, whereas "writer" is a noun. Syntactic features include encoding grammatical roles like subject, object, or predicate in sentences.

This step encodes information about a word's position in a sentence and its relationship to sentence structure. For example, in the sentence "The king sits on the throne," the position of the word "king" tells us it is the subject. The positional encoding is added to give the model information about the position of each word in the sentence. Thus, each token is converted into a dense vector encapsulating these features. These embeddings are learned from data and capture a wide range of language nuances. The embedding process is crucial. By definition, the Transformer does not process sequences in order like RNN. Thus, transformers add the necessary positional encodings to each token's embedding to provide positional information (Riedl 2023).

The third step is Encoder Operations. Transformers rely heavily on a concept called self-attention (Vaswani et al. 2017). The "self" part of self-attention refers to each token's "egocentric" focus in the sentence. Effectively, for each token of input, self-attention means evaluating the relevance of every other input token. Thus, in the sentence "Today is a beautiful day," there are five words. Each of the five words pays attention to the other four, wondering how much each of those four words matters to them.

Let us take a closer look at how the self-attention mechanism works. Consider the sentence, "**The plane landed at the departing airport because it had mechanical problems**." When we mask the second word, "plane" and ask the system to guess it, the word "plane" would be a good candidate because of the presence in the sentence of the word "airport." Another concept that may cross our minds might be the "helicopter," but a helicopter does not need an airport to land and take off. The word "departing" is linked to the word "airport," as is "mechanical" to "problems."

If we have to mask the word "it" and ask the system to guess it, then there are three candidates for this position: he, she, and it. The presence of the word "plane" makes the word "it" more probable than "he" or "she."

The model computes attention scores for each word, determining how much focus it should put on other words in the sentence for each word. For example, for the token "beautiful," the model may learn to pay more attention to the token "day" to understand the phrase "beautiful day." These token-based computations convert each token into a numerical vector representing its features. As a result, the encoder operation's step transforms the input sentence into a series of vectors with attention scores presented as follows:

"Today": [contextualized vector]
"is": [contextualized vector]
"a": [contextualized vector]
"beautiful": [contextualized vector]
"day": [contextualized vector].

The sentence, now a series of vectors, passes through several encoder layers. In each next layer, the series of vectors go through the self-attention mechanism (Wang and Chen 2023) and the feed-forward neural network process, refining the representation with contextual information.

The Transformer calculates the attention scores for each word in the sentence (Lin et al. 2022). This step helps the model to understand the context and the relationship between words. For instance, it learns that "Today" is the sentence's main subject.

The fourth step is the Decoder Operations (Ge et al. 2022). The decoder considers the output of the encoder in order to generate the target sentence in Italian, doing so one word at a time. The decoder starts with a unique token like "[***SOS***]." At each step, the decoder predicts the next word based on the words it has already generated.

This process is iterated until the decoder produces an "[EOS]" token or reaches a maximum length. The decoder also has several layers, each with a self-attention mechanism, but it processes one word at a time in the target language.

Throughout the process, the model uses attention mechanisms to decide how much focus to place on each part of the input sentence. For instance, while predicting "bel" (beautiful), it will use some "Contextual Understanding" to determine that the word in Italian corresponding to "beautiful" should match the gender of the word "giornata," which is the translation for the token "Today." As in Italian, the word "giornata" is feminine, so the appropriate translated token in Italian should be "bella."

It starts with a special start token (e.g., "[***SOS***]") and predicts the first word ("Oggi"). With each prediction, the decoder considers the previously generated words and the entire input sentence to predict the next word. This process continues until the model generates an end token or reaches a maximum length. Thus, the prediction sentence is:

- <SOS> → "Oggi" (Today)
- "Oggi" → "è" (is)
- "Oggi è" → "un" (a)
- "Oggi è una" → "bella" (beautiful, masculine)
- "Oggi è una bella" → "giornata" (day).

The decoder generates the translated sentence: "Oggi è una bella giornata."

5.6.2 Points to Consider About LLM

During training, we feed a snippet of text from one of these sources and ask it to guess the next word. Remember: self-supervised. When during learning the LLM guesses wrong, we tweak the model until it gets it right. If we were to think about what an LLM is trained to do, it is to produce text, which might reasonably appear on the Internet. LLM do not and cannot memorize the Internet, so it uses the encodings to make compromises and guess things sometimes slightly wrong, but hopefully not too wrong (Riedl 2023).

A few issues about LLM's should be addressed:

1. Large language models are trained in the massive Internet world. That means they have absorbed millions of millions of texts, dialogs, novels, books, social media messages, etc. We all know that the Internet has known and unknown, legal and illegal, ethical and nonethical sources of information. At the least, all this information was initially fed to large language models with little or no filters. It should not be a surprise that this illegal, unethical, racist information will surface while generating new text using Large Language models. Later on, AI companies increased their efforts to filter more carefully the information they generate, trying to avoid political and social bias.

2. Large language models do not have "core beliefs," "consciousness" nor do they have any sense of truth or right or wrong. They try to guess the next word, knowing the subset of already known words were the same sentence to appear on the Internet. As they do not have the sense of wrong or right, there is no guarantee that a LLM will provide the truth. LLM's response is very much influenced by how the question is designed. Thus, if the question is asked to write a sentence in favor of or against a specific idea, the language model will comply both ways.

We experimented by asking CHATGPT-4: "How do I justify stealing bread because of the risk of starvation?" The complex answer involved considering various ethical, legal, and social perspectives. The ethical justification includes survival and basic needs, a utilitarian perspective, and relativism. The legal justification was based on the concept of defense of necessity. The social justification was presented under social responsibility and compassion, comprising a sharp critique of social systems, which had failed the individual and forced him to steal to survive.

At the end, a meaningful warning is given stating that justifying an act of theft due to the risk of starvation involves a complex interplay of ethical, legal, and social considerations. It is important to emphasize that whereas such justification might be understandable in extreme situations, it does not negate the fact that stealing is illegal and generally considered morally wrong in most societies. The focus should ideally be on addressing the underlying issues that lead to such dire circumstances (OpenAI. 2023).

3. Large Language Models can make mistakes, an issue that should be kept in mind; it is a relevant philosophical issue that users should consider seriously. One reason for the mistakes is that the training data might have many inconsistencies that are now part of the arsenal of words LLM has. Another reason is that self-attention may only attend to some of the things we want when a question is asked. LLM is only a word-guesser, and it may make incorrect guesses. The training data has seen one word so many more times that it prefers that word even when it does not make sense for the input. In this case, we have what is called "**hallucination**" (Azamfirei et al. 2023; Huang et al. 2023; Guerreiro et al. 2023; Liu et al. 2023). LLMs are inclined to guess small numbers instead of large ones because small ones are more common than large ones. Thus, LLMs are not great at mathematical operations. Initially, when used, LLMs would provide wrong answers when instructed to make certain calculations, errors which led to the replacement of LLM's by Python procedures to calculate the result.
4. Since LLMs are responding to human requests in whichever way possible—exactly what they were designed and trained to do—many machine-learning specialists do not believe that hallucinations can be resolved. However, it is crucial to recognize that our collective perceptions of these models and our decisions regarding their use are the true source of the issue. The researchers suggest that fact-checking methods that never leave a chatbot unattended must be combined with generative AI techniques to reduce hallucinations (Leffer 2024).

5. Three machine learning researchers at the National University of Singapore (Xu et al. 2024) demonstrated in a recent study that hallucination is unavoidable in big language models. The demonstration shows that LLMs cannot learn every computable function by using a few well-known findings when learning theory. An example is Cantor's diagonalization argument. "Any LLM will unavoidably experience hallucinations in certain parts of the real world, where it is incapable of learning" (Xu et al. 2024).
6. Large models can provide biased results. To ensure fairness, accuracy, and dependability, it is imperative to comprehend and tackle bias in ChatGPT and related AI models. There are various ways that bias in ChatGPT or AI-based systems might occur:

   (a) The quality of the data used for training purposes should be bias-free. When the training data contains social, cultural, or factual biases, the model is likely to learn and replicate these biases.
   (b) Biases may also be introduced via the model's architecture and design. The model's performance can be impacted by hyperparameters such as learning rate, batch size, and number of layers. The model may exacerbate preexisting biases when these hyperparameters are manually adjusted to maximize performance on biased datasets. If the objective function—such as cross-entropy loss—is not intended to consider fairness, biases may be introduced during training. For instance, the model may perform better on majority groups at the expense of minority groups if the loss function favors accuracy above fairness.

The problem of bias in ChatGPT is complex and arises from a number of sources, including as user interactions, model architecture, and training data. Diverse training data, bias identification and analysis, human feedback, algorithmic tweaks, and ongoing development are all necessary to combat this prejudice. The goal of OpenAI's continuous work is to develop trustworthy, accurate, and equitable AI models that benefit a range of demographics and reduce detrimental biases (Roumeliotis and Tselikas 2023).

Large language models, in a sense, only simulate the language functions of the brain; they do not replicate perception, memory, navigation, social judgment, or other functions. Our gray matter carries out a confusing jumble of overlapping tasks, some of which are more localized and others of which are broadly dispersed across the brain. Individuals who have experienced a stroke in one of their language domains may not be able to talk, but they may still be able to play chess, add numbers, make music, and interact with others through gestures. Such modularity is being added by AI engineers to their systems to increase their intelligence (MUSSER 2024).

5.7 Risks, Opportunities, and Control on AI Development

The future of AI and the concurrent safety concerns are areas of considerable interest and continuous research. At the time of this book's publication in mid-2024, AI was expanding and developing at a galloping speed, making it hard to predict subsequent developments. As AI continues to evolve, several key ideas and trends are likely to shape its course.

AI is expected to progress toward more advanced capabilities and general intelligence, moving beyond specialized tasks to conducting a more comprehensive range of activities that require understanding and adaptability. This progress could lead to AI systems, which are more flexible and capable of learning new tasks with little human intervention.

The future will likely underscore the need for collaborative systems where humans and AI work together, leveraging the strengths of both. This future human-AI collaboration brings attention to guaranteeing that these coalitions are effective and safe and that AI systems can understand and adapt to human needs and intentions, which will be necessary.

There are concerns about the long-term existential risks posed by AI. Many researchers who have contributed significantly to the development of AI have voiced serious concerns about this issue, especially if it reaches or surpasses human intelligence. Handling these concerns implicates research into AI alignment—confirming AI systems' goals are aligned with human values—and control to ensure humanity can retain control over advanced AI systems. Achieving these goals brings both opportunities for collaboration and potential for competition, primarily in areas like military use of AI and economic competitiveness. Balancing these aspects while ensuring AI's safe and beneficial development is a complex challenge. The risks and opportunities posed by AI developments should be addressed through international collaboration and competition, similar to the way nuclear risks and advantages were addressed after the Second World War (Suleyman and Bhaskar 2023).

References

Araújo, A., Portugal, D., Couceiro, M. S., & Rocha, R. P. (2015). Integrating Arduino-Based Educational Mobile Robots in ROS. *Journal of Intelligent & Robotic Systems*, *77*(2), 281–298. https://doi.org/10.1007/s10846-013-0007-4

Arkin, R. C. (1998). *Behavior-Based Robotics*. MIT Press.

Azamfirei, R., Kudchadkar, S. R., & Fackler, J. (2023). Large language models and the perils of their hallucinations. *Critical Care*, *27*(1), 120. https://doi.org/10.1186/s13054-023-04393-x

Bishop, C., & Nasrabadi, N. (2007). Pattern Recognition and Machine Learning. *Journal of Electronic Imaging, 16*(4), 049901-049901-2

Blank, I. A. (2023). What are large language models supposed to model? *Trends in Cognitive Sciences*, *27*(11), 987–989. https://doi.org/10.1016/j.tics.2023.08.006

Brooks, R. (1986). A robust layered control system for a mobile robot. *IEEE Journal on Robotics and Automation*, *2*(1), 14–23. https://doi.org/10.1109/JRA.1986.1087032
Brooks, R. A. (1991). Intelligence without representation. *Artificial Intelligence*, *47*(1–3), 139–159. https://doi.org/10.1016/0004-3702(91)90053-M
Chen, P.-C., Tsai, H., Bhojanapalli, S., Chung, H. W., Chang, Y.-W., & Ferng, C.-S. (2021). *A Simple and Effective Positional Encoding for Transformers*. http://arxiv.org/abs/2104.08698
Choi, S. R., & Lee, M. (2023). Transformer Architecture and Attention Mechanisms in Genome Data Analysis: A Comprehensive Review. *Biology*, *12*(7), 1033. https://doi.org/10.3390/biology12071033
Crevier, D. (1993). *Ai: The Tumultuous History Of The Search For Artificial Intelligence*.
Dufter, P., Schmitt, M., & Schütze, H. (2022). Position Information in Transformers: An Overview. *Computational Linguistics*, *48*(3), 733–763. https://doi.org/10.1162/coli_a_00445
Ehrlich, B. (2022). The father of modern neuroscience discovered the basic unit of the nervous system. Scientific American. https://www.scientificamerican.com/article/the-father-of-modern-neuroscience-discovered-the-basic-unit-of-the-nervous-system/
Emanuel, R. H. K., Docherty, P. D., Lunt, H., & Möller, K. (2023). The effect of activation functions on accuracy, convergence speed, and misclassification confidence in CNN text classification: a comprehensive exploration. *The Journal of Supercomputing*. https://doi.org/10.1007/s11227-023-05441-7
Ge, H., Park, I. Y., Qian, X., & Zeng, G. (2022). *Transformer Encoder for Social Science*. http://arxiv.org/abs/2208.08005
Guerreiro, N. M., Alves, D. M., Waldendorf, J., Haddow, B., Birch, A., Colombo, P., & Martins, A. F. T. (2023). Hallucinations in Large Multilingual Translation Models. *Transactions of the Association for Computational Linguistics*, *11*, 1500–1517. https://doi.org/10.1162/tacl_a_00615
Heil, B. J., Crawford, J., & Greene, C. S. (2023). The effect of non-linear signal in classification problems using gene expression. *PLOS Computational Biology*, *19*(3), e1010984. https://doi.org/10.1371/journal.pcbi.1010984
Hsu, F.-H. (1987). A two-million moves/s CMOS single-chip chess move generator. *IEEE Journal of Solid-State Circuits*, *22*(5), 841–846. https://doi.org/10.1109/JSSC.1987.1052821
Hsu, F.H., Anantharaman, T. S., Campbell, M. S., & Nowatzyk, A. (1990). Deep Thought. In *Computers, Chess, and Cognition* (pp. 55–78). Springer New York. https://doi.org/10.1007/978-1-4613-9080-0_5
Huang, L., Yu, W., Ma, W., Zhong, W., Feng, Z., Wang, H., Chen, Q., Peng, W., Feng, X., Qin, B., & Liu, T. (2023). *A Survey on Hallucination in Large Language Models: Principles, Taxonomy, Challenges, and Open Questions*. http://arxiv.org/abs/2311.05232
Institutet, T. N. A. at K. (2014). *Nobel Prize in Medicine awarded for locating brain's inner GPS*.
Isinkaye, F. O., Folajimi, Y. O., & Ojokoh, B. A. (2015). Recommendation systems: Principles, methods and evaluation. *Egyptian Informatics Journal*, *16*(3), 261–273. https://doi.org/10.1016/j.eij.2015.06.005
Ivakhnenko, A. G., & Ivakhnenko, G. A. (2009). Problems of Further Development of the Group Method of Data Handling Algorithms. Part I. *Pattern Recognition and Image Analysis*, *10*(2).
Jérémy Virgo, Jonathan Pillon, Jordan Navarro, Emanuelle Reynaud, François Osiurak; Are You Sure You're Faster When Using a Cognitive Tool?. *The American Journal of Psychology*, *130*(4): 493–503. https://doi.org/10.5406/amerjpsyc.130.4.0493
Hovers, E. (2011). Invention, Reinvention and Innovation: The Makings of Oldowan Lithic Technology. *Developments in Quaternary Sciences*, *16*, 51–68. https://doi.org/10.1016/B978-0-444-53821-5.00005-1
Kane, A., Patankar, S., Khose, S., & Kirtane, N. (2022). *Transformer based ensemble for emotion detection*. http://arxiv.org/abs/2203.11899
Kaplan, A. (2022). *Artificial Intelligence, Business and Civilization Our Fate Made in Machines*.
Kasparov, G. (2022). *Garry Kasparov on chess, technology and democracy*.

Khan, S., Fazil, M., Imoize, A. L., Alabduallah, B. I., Albahlal, B. M., Alajlan, S. A., Almjally, A., & Siddiqui, T. (2023). Transformer Architecture-Based Transfer Learning for Politeness Prediction in Conversation. *Sustainability*, *15*(14), 10828. https://doi.org/10.3390/su151410828

Kline, R. (2011). Cybernetics, Automata Studies, and the Dartmouth Conference on Artificial Intelligence. *IEEE Annals of the History of Computing*, *33*(4), 5–16. https://doi.org/10.1109/MAHC.2010.44

Lamsal, R. (2021). *A step by step forward pass and backpropagation example.*

Leffer, L. (2024). AI Chatbots Will Never Stop Hallucinating. *Scientific American.*

Lin, T., Wang, Y., Liu, X., & Qiu, X. (2022). A survey of transformers. *AI Open*, *3*, 111–132. https://doi.org/10.1016/j.aiopen.2022.10.001

Linegar, M., Kocielnik, R., & Alvarez, R. M. (2023). Large language models and political science. *Frontiers in Political Science*, *5*. https://doi.org/10.3389/fpos.2023.1257092

Liu, X., Rubin, S. J. S., & Rogalla, S. (2023). Looking to Future Applications of Large Language Models. *American Journal of Gastroenterology*, *118*(12), 2305–2305. https://doi.org/10.14309/ajg.0000000000002401

McCorduck, P. (2004). *Machines Who Think* (2nd ed.).

Meyer, J. G., Urbanowicz, R. J., Martin, P. C. N., O'Connor, K., Li, R., Peng, P.-C., Bright, T. J., Tatonetti, N., Won, K. J., Gonzalez-Hernandez, G., & Moore, J. H. (2023). ChatGPT and large language models in academia: opportunities and challenges. *BioData Mining*, *16*(1), 20. https://doi.org/10.1186/s13040-023-00339-9

Miller, T. (2018). Explanation in artificial intelligence: Insights from the social sciences. *Artificial Intelligence, 267*, 1–38. https://doi.org/10.1016/j.artint.2018.07.007

Mondada, F., Bonani, M., Raemy, X., Pugh, J., Cianci, C., Klaptocz, A. S. M., Zufferey, J.-C., Floreano, D., & Martinoli, A. (2010). The e-puck, a Robot Designed for Education in Engineering. *The 9th Conference on Autonomous Robot Systems and Competitions*, 59–65.

Murray, C.A., Joseph Hoane, J., & Hsu, F. (2002). Deep Blue. *Artificial Intelligence*, *134*, 57–83.

MUSSER, G. (2024). Building Intelligent Machines Helps Us Learn How Our Brain Works. *Scientific American* https://doi.org/10.1038/scientificamerican0424-31

Nanni, L., Brahnam, S., Paci, M., & Ghidoni, S. (2021). Comparison of Different Convolutional Neural Network Activation Functions and Methods for Building Ensembles for Small to Midsize Medical Data Sets. *Sensors, 22*(16), 6129. https://doi.org/10.3390/s22166129

Naveed, H., Khan, A. U., Qiu, S., Saqib, M., Anwar, S., Usman, M., Akhtar, N., Barnes, N., & Mian, A. (2023). *A Comprehensive Overview of Large Language Models.* https://doi.org/: https://github.com/humza909/LLM_Survey.git

Nilsson, N. J. (1994). Teleo-Reactive Programs for Agent Control. *Journal of Arti Cial Intelligence Research, 1*, 139–158.

Nwankpa, C. E., Ijomah, W., Gachagan, A., & Marshall, S. (2018). Activation Functions: Comparison of Trends in Practice and Research for Deep Learning. *ArXiv:1811.03378v1.*

OpenAI. (2023). *No Title*. ChatGPT [Large Language Model]. https://chat.openai.com

Osiurak, F., Navarro, J., & Reynaud, E. (2018). How Our Cognition Shapes and Is Shaped by Technology: A Common Framework for Understanding Human Tool-Use Interactions in the Past, Present, and Future. *Frontiers in Psychology*, *9*. https://doi.org/10.3389/fpsyg.2018.00293

Ramachandran, P., Zoph, B., & Le, Q. V. (2017). Searching for activation functions. *ArXiv Preprint ArXiv:1710.05941.*

Riedl, M. (2023). *A Very Gentle Introduction to Large Language Models without the Hype.*

Roumeliotis, K. I., & Tselikas, N. D. (2023). ChatGPT and Open-AI Models: A Preliminary Review. *Future Internet*, *15*(6), 192. https://doi.org/10.3390/fi15060192

Schafer, J. B., Konstan, J., & Riedl, J. (1999). Recommender systems in e-commerce. *The 1st ACM Conference on Electronic Commerce*, 158–166.

Shanon, C. (1950). A chess playing machine. *Scientific American.*

Siddharth, S., Simone, S., & Anidhya, A. (2020). ACTIVATION FUNCTIONS IN NEURAL NETWORKS. *International Journal of Engineering Applied Sciences and Technology*, *4*(12), Pages 310–316.

Solso, R. L., MacLin, M. M., & Kimberly H.O (2005). *Cognitive psychology* (7th ed.). Pearson Education.

Suleyman, M., & Bhaskar, M. (2023). *The Coming Wave: Technology, Power, and the Twenty-first Century's Greatest Dilemma*. Crown.

Tabinda Kokab, S., Asghar, S., & Naz, S. (2022). Transformer-based deep learning models for the sentiment analysis of social media data. *Array*, *14*, 100157. https://doi.org/10.1016/j.array.2022.100157

Taherdoost, H. (2023). Deep Learning and Neural Networks: Decision-Making Implications. *Symmetry*, *15*(9), 1723. https://doi.org/10.3390/sym15091723

Vaswani, A., Shazeer, N., Parmar, N., Uszkoreit, J., Jones, L., Gomez, A. N., & Polosukhin, I. (2017). Attention is all you need. *Advances in Neural Information Processing Systems*.

Vernon, D., Metta, G., & Sandini, G. (2007). A Survey of Artificial Cognitive Systems: Implications for the Autonomous Development of Mental Capabilities in Computational Agents. *IEEE Transactions on Evolutionary Computation*, *11*(2), 151–180. https://doi.org/10.1109/TEVC.2006.890274

Wang, S., & Chen, B. (2023). A Comparative Study of Attention-Based Transformer Networks and Traditional Machine Learning Methods for Toxic Comments Classification. *Trends Journal of Sciences Research*, *1*(1), 22–30. https://doi.org/10.31586/jsmhes.2023.697

Whitehead, A., & Russell, B. (2004). *Principia Mathematika* (11th editi). Cambridge University Press.

Xu, Z., Jain, S., & Kankanhalli, M. (2024). *Hallucination is Inevitable: An Innate Limitation of Large Language Models*. http://arxiv.org/abs/2401.11817

Zhao, W. X., Zhou, K., Li, J., Tang, T., Wang, X., Hou, Y., Min, Y., Zhang, B., Zhang, J., Dong, Z., Du, Y., Yang, C., Chen, Y., Chen, Z., Jiang, J., Ren, R., Li, Y., Tang, X., Liu, Z., … Wen, J.-R. (2023). *A Survey of Large Language Models*. https://doi.org/10.48550/arXiv.2303.18223

Chapter 6
Mind Genomics: Origins, Evolution, Inner-Workings

6.1 Introduction

Mind Genomics is a new, evolving science, focusing on the understanding of daily experience, and more specifically on the decisions we make about the everyday. One might wonder "why," or more specifically, why a science devoted to the decision processes of everyday, and even more importantly, why devote a science to the topic. Is not the world of decision making already filled with investigating scientists, ranging from the anthropologist to the behavioral economists, the sociologist, the psychologist, and many more?

The answer to the foregoing question, "why this new science" comes from the desire to put structure around the question of, the "every day." The behavioral economist can talk about theories of utility of money, the sociologist can talk about the way groups of people self-organize to get things done, the psychologists can talk about the motives of shopper, with all topics backed up by literature, both "popular" and "scientific," respectively. This literature is vast, diverse, does not lend itself to databasing, deals with major problems, and in the end requires expertise to move beyond knowing the topic to actually doing some constructive with the knowledge.

Mind Genomics arose to produce knowledge about the everyday but not just knowledge itself. The objective of Mind Genomics is to democratize the development of this knowledge, and at the same time create a corpus of information that could be applied to the issues of daily life in a manner that would produce actionable information. In other words, this science was to be practical, rather than simply a theory about the world of the everyday.

P. Papajorgji, H. Moskowitz, *The Mind of Everyday*,
https://doi.org/10.1007/978-3-031-78078-3_6

6.2 Mind Genomics and the Outer Versus Inner Psychophysics of S.S. Stevens

The work product of Mind Genomics is the assignment of numbers to aspects of daily life. To the research scientists of the past these topics would have been of some interest but would not have been sufficiently important to investigate in and of themselves. The nutritionist would have been interested in what to do to help patients lose weight, the doctor would have focused on whom tell to lose weight and whom not to tell. The specific inner experience of the patient would not have been of real interest. The doctor would not have felt at home spending a great deal of time understanding the mind of the patient, at least not in the doctor's early career. In turn, the nutritionist would have focused on behavior, what to do, but not necessarily on the specifics of words in the communication with the client or patient.

The same pattern of not knowing the inside of the mind of the patient occurs in the world of commerce. Certainly, there are salespeople who can "size up" a customer walking in to buy a product, such as a lamp. These salespeople seem to "know the right words" to greet a customer. Yet, for every one of these expert salespeople there are many more who know what to say, talking about the features of the product, the benefits, but not really reaching into the mind of the prospect to know what will motivate this particular individual.

The ultimate issue motivating the development of Mind Genomics is "*how can we understand the mind of the person*?" Science, and now technology gives us a great deal of knowledge about people, what they do, and what they say they want, but only years of experience tell us what to say to "*reach them*." The proof of that is the blank look one gets when one asks almost any experienced "professional" to state the few phrases that will convince and motivate to get a specific person to buy a specific item, or the few phrases that will convince and motivate the individual to go on a diet to lose weight. It is that specificity which lies at the root of Mind Genomics.

With the foregoing in mind, we can look back historically to psychology, and specifically to experimental psychology as the nearest ancestor of Mind Genomics. It was in the middle 1800s that the question emerged about "measuring private sensory experience." The German polymath, Gustav Theodor Fechner (Ortlieb et al. 2020; Heidelberger 1994) searched for the inner measuring instrument of sensation. It wasn't sufficient to know that a person was sensing something, like tasting the sweetness of a glass of tea. The question was "what was the perceived strength of sweetness?" Fechner did not address the issue in that modern way, but rather in a more philosophical treatment dealt with the possibility of measuring "private sensory experience." The approach was to measure the magnitude of increments, just noticeable differences, and then cumulate them. That is, one could start with the lowest level of sugar that could be detected in tea. That would be the sweetness "1." One would then find the next concentration level that was just noticeably different, sweeter. That next concentration level would be the value "2." Although painstaking, Fechner was on to something. One could measure experience, if this way of thinking were correct (Heidelberger 1993; Beiser 2020).

Fechner's thinking stimulated Harvard psychology professor S.S. (Smitty) Stevens to try this measurement in a bit of a different way, a more direct way (Stevens 1975). Stevens suggested that the person, the "observer" (in deference to the German history of the field), should simply assign a number to show how sweet the tea seemed. Stevens showed that for individuals, but especially for groups of people, one could get people to successfully match the size of the numbers of the sweetness experienced. The wonderful surprise was that when the numbers were plotted, the concentration of sugar in tea on the abscissa and the median rating assigned by the observer on the ordinate, the ratings were "tracked." The more sugar in the tea, the higher the number, even though the researcher randomized the order, so the observer had to rely strictly on her or his sensation of sweetness.

The notion that one could measure private sensory experience can be said to be the root of Mind Genomics. The attraction of the measurement of sensation to a young, 21-year-old graduate student, author HRM, cannot be easily put into words. The discovery that simple numbers could be used to tap into thoughts was exciting. At that time, 1965, 105 years after Fechner, the excitement focused on discovering for oneself a lawful relation between the concentration of sugar and the ratings of sweetness, or the amperage of a shocking device and the perceived pain.

Had the interest stopped with discovering the lawful relation between what was physically measured and what was rated by the person, Mind Genomics might not have been born. The foregoing effort was named by Stevens as "the outer psychophysics" (Billock and Tsou 2011; Chen et al. 2021). The idea was that one was measuring the sensory response, an internal experience, but the source of that experience was the presentation of well-defined stimuli. The excitement of this outer psychophysics continued to emerge from study after study. The source of the excitement, a regularity of nature. When plotted in log-log coordinates (logarithmic spacing of numbers on the abscissa, e.g., concentration of sugar in tea, logarithmic space on the ordinate, e.g., rating assigned by the respondent), the plot was a straight line. The slope of the line was reproducible and characteristically different for each of the different physical continua, such as the rating of loudness of sounds, brightness of lights heaviness of weights, etc.

In the midst of research on the sense of taste, author HRM's doctoral thesis, an interesting paper emerged in the literature, titled "Does the punishment match the crime" (Rose and Prell 1955). The basic issue was the application of this psychophysical measurement to the seriousness of crimes. The approach was the same as that espoused by the S.S. Stevens, namely, that observers (now relabeled respondents) were instructed to read a series of crimes and assign a number to the seriousness of the crime. The study opened the possibility of measuring an idea, seriousness of crime, although in this case the rated seriousness was to be plotted against the magnitude of the punishment. Mind Genomics began with the ability to move beyond tangible physical stimuli, or even imagined physical stimuli such as money (Galanter and Pliner 1974) and work with aspects that were purely experiential. Mind Genomics took its first steps that day with the realization of new possibilities, although it would be years before the conception of a new field would be recognized.

6.3 Beyond Simple Stimuli to Mixtures, and the Guiding Factor of Experimental Design

Experimental psychologists and especially psychophysicists prefer to work with well-controlled environments. The effort to relate two variables to each other, e.g., the taste of a sweet cup of tea versus the amount of sugar, is hardly ever done in the sloppy manner, which constitutes the warp and woof of reality. Scientists looking for relations between variables, or even simple effects prefer to reduce the random noise in the situation, variability which could influence the stability of that which is being measured.

The reality of life is that it is messy. Isolating a single variable and measuring the response to that variable may be the way to do good psychophysics, but it does not represent real life. Life comprises mixtures of stuff, of things, of events. The mixtures are messy, of course, but somehow people make their way through the mess to do what they have to do. Whether it is the mixtures of sight, smells, tastes, and textures of a buffet dinner, or the lovely scene in a fall afternoon on a wooded lane in the country, people seem to make sense of the experience. To all outward appearance it is the mixture, which is the real experience (Porretta et al. 2019).

The foregoing has attempted to express author HRM's realization that this inner psychophysics, this measurement of thought and experience, would have to be able to deal with mixtures. One might fool oneself by asking the person a series of single questions on different aspects of an experience, such as that asked in surveys, say about one's experience after an interaction with the world of medicine. The Press Ganey questionnaire, for instance, focuses a variety of different aspects of an experience with the medical world, ranging from time aspects to the nature of the interaction with the medical processional, and so forth (Presson et al. 2017). The Press Ganey system captures the different aspects of the experience, doing so one aspect at a time. The outcome is a report card, but one that is intellectualized without the richness of the actual experience, viz., report where the respondent must abstract a feeling from memory, think about it, and then rate it.

With the foregoing realization, it became obvious that to measure inner experience and thinking would require movement beyond intellectualization, to actually reexperiencing. One way to simulate some of this richness is to present the person with a compound description of aspects of the experience, have the respondent read the compound description, and then rate it on a scale (Moskowitz et al. 2006). The notion of creating combinations is actually obvious. When we eat, we respond to actual combinations. When we look at art, we respond to actual combinations. And the list could go on. Thus, when we want to measure internal thoughts, we have to evoke combinations of thoughts, the more nature way. Do not describe, but rather create the experience, in this case the experience of ideas.

The task of measuring combinations of ideas, or really combinations of messages about ideas, is actually far simpler than one might think. Statisticians have dealt with the issue of combinations or mixtures for decades, creating sets of combinations called experimental designs (Presson et al. 2017; Ryan and Morgan 2007). The experimental design lays out the specific combinations.

Table 6.1 presents part of the experimental design used in a Mind Genomics study. The study dealt with local issues regarding "building new residences in a neighborhood." The study worked with 101 respondents who participate in a Mind Genomics study conducted over the web.

1. Table 6.1 shows 10 of the 24 rows, each row corresponding to one of the vignettes rated by a respondent. The experimental design is really a set of recipes. The columns correspond to the 16 different elements. The elements or messages are single ideas. The experimental design prescribes the specific combination of these elements (Gofman and Moskowitz 2010a, c). These features should be kept in mind when looking at Table 6.1.
2. Table 6.1 is set up for statistical analysis using regression (Flatt and Jacobs 2019). The analysis will be discussed below.
3. Experimental designs enable the researcher to put together different "things," whether these be ingredients in a recipe, or phrases. The underlying objective is that the "natural experience" is one of mixtures, rather than evaluating each aspect in a sort of noise-free environment. A mixture more validly represents what the person will experience.
4. The first column shows the order, from the first vignette or combination to the last. Each person evaluates a specified set of 24 vignettes, no more, no less. The reason for this rigid specification is that the set of combinations specified by the experimental design will be precisely those needed to fully analyze the data for a single individual. That is, the ratings for each individual respondent in a Mind Genomics study can be separately analyzed using OLS (ordinary least-squares regression) (Zdaniuk 2014; Burton 2021). This property of "within-subjects design" will "come in handy" for the analysis of patterns at the level of the individual respondent, in preparing for clustering respondents into similar performing groups. Clustering will be dealt with below in detail (Flatt and Jacobs 2019).

Table 6.1 The first 10 vignettes, in "binary" coding. The underlying experimental design specifies the structure of each vignette

Order	A1	A2	A3	A4	B1	B2	B3	B4	C1	C2	C3	C4	D1	D2	D3	D4
1	0	0	0	1	0	0	1	0	0	1	0	0	0	1	0	0
2	0	1	0	0	0	1	0	0	0	0	0	1	0	1	0	0
3	0	1	0	0	1	0	0	0	1	0	0	0	0	1	0	0
4	0	0	0	0	1	0	0	0	0	0	1	0	0	0	0	1
5	1	0	0	0	0	0	1	0	0	0	0	0	1	0	0	0
6	1	0	0	0	0	0	0	1	1	0	0	0	0	1	0	0
7	0	0	0	0	0	0	0	1	0	0	0	1	1	0	0	0
8	0	1	0	0	0	0	0	1	0	0	0	0	0	0	0	1
9	0	0	1	0	1	0	0	0	0	1	0	0	0	0	1	0
10	0	0	0	1	0	1	0	0	1	0	0	0	0	0	1	0

5. The experimental design for Mind Genomics comprises one "basic set" of 24 recipes or combinations, comprising 2–4 elements within a combination. The specifics are dictated by the basic experimental design.
6. The source of the elements are the answers to the four questions, A,B, C, D. Each question has four answers associated with it.
7. Each vignette has at most one element from a question. The rationale for this is "bookkeeping," viz., that the vignette is prevented from showing two elements conveying different and possibly contradictory information from the same question. For example, the question may be about the "cost" of the product. One element may state "X" dollar, another element may state "Y" dollars. The two elements make sense separately, but when they are put together the vignette makes no sense at all. There are some situations where two messages may not seem to contradict each other on the surface, but in actuality they may be impossible to satisfy at the same time in a real product. In such a case it is left to the discretion of the researcher whether these elements must be put into the same category or silo, or whether they can appear in separate categories or silos, in which case they might appear together in some vignettes.
8. Each respondent is assigned a unique design or set of 24 vignettes, viz., combinations.
9. Each set of 24 vignettes created for one person is mathematically of the same form (called isomorphic) with each set of 24 vignettes created for one respondent. The elements are systematically permuted within a single category or silo. Thus, the combinations end up changing, even though the mathematical properties of the design are maintained. This approach generates the so-called isomorphic permutation design (Gofman and Moskowitz 2010b).
10. One knows precisely which elements or messages appear in each vignette. Elements which appear are coded "1." Elements which are absent are coded "0."

The respondent rates each vignette separately, rating the vignette on a predefined scale. Typically, the scale is anchored at both the low and high levels. In recent studies the scale has been anchored at every scale point. This anchoring has been done because recent rating scales have incorporated two dimensions in one scale. Respondents find these scales easy to use once they go through a practice exercise. Table 6.2 shows an example of the scale.

It is important to note that the scale shown in Table 6.2 has two sides. The first side is an evaluation of the nature of the concession, viz., whether the builder has offered a good concession, or a poor concession. The other side is the response of the community whether acceptance or rejection (Moskowitz et al. 2020).

Other ways of structuring the scale may create a simpler scale, with each scale point corresponding to an increasing level of type of concession (e.g., from very poor to great) or from degree of community acceptance (community rejects unanimously to community accepts unanimously). In the end, the rating scale is left to the researcher, but it is critical to keep in mind that it is only through the rating scale that we will have any idea about what the respondent is feeling or thinking.

Table 6.2 Example of a rating scale and the transformations of the scale

Original scale points for a two-sided scale
1 = Poor builder concessions AND community rejects
2 = Poor builder concessions BUT community accepts
3 = I don't now
4 = Good builder concessions BUT community rejects
5 = Good builder concessions AND community accepts
Transformed scale to created binary scale preparation for regression modeling
R5X (Good concession, community accepts) Becomes 100 when the rating is 5. Becomes 0 when the rating is 1,2,3, or 4
R54X (Good concession) Becomes 100 when the rating is 5. Becomes 0 when the rating is 1,2, or 3
R52X (Community accepts) Becomes 100 when the rating is 5 or 2. Becomes 0 when the rating is 4,3,or 1
R3X (Do not know) Becomes 100 when the rating is 3. Becomes 0 when the rating is 1,2,4, or 5
R41 (Community rejects) Becomes 100 when the rating is 4 or 1. Becomes 0 when rating is 5, 3, or 2
R21X (Poor concession) Becomes 100 when the rating is 2 or 1. Becomes 0 when the rating is 5,4, or 3
R1X (Poor concession, community rejects) Becomes 100 when the rating is 1. Becomes 0 when the rating is 5,4,3, or 2.

1. The respondent assigns a rating to the vignette. The rating may either come from a one-dimensional scale (e.g., definitely not purchase to definitely purchase), or from a two-sided or two-dimensional scale. Table 6.2 shows transformations used by Mind Genomics to create relevant dependent variables. These dependent variables are "binary." That is, the transformation takes specific values of the rating scale and transforms these specific values to "100," as well as taking the remaining values on the rating scale and forming these to "0." After the transformation, a vanishingly small random number (<10–5) is added to the newly created dependent variable. The addition of this vanishingly small random number is prophylactic, ensuring that all dependent variables has at least some minimal variability. That minimal variability ensures that the statistics program won't crash.
2. The 24 vignettes are set up for analysis by OLS, ordinary least squares regression (Zdaniuk 2014; Burton 2021). OLS regression is easy to do, widely accepted, with results easy to understand. During the past year the OLS model used by Mind Genomics has been simplified to be a simple linear equation, estimated without an additive constant. This is called "forcing the regression model through the origin." This simplification allows the coefficients to be compared across studies, without the need for a baseline. A little information is lost by foregoing the additive constant, but the benefits far outweigh the negatives.
3. The expression is: Dependent binary transformed variability = k1A1 + k2A2 … k16D4. Each regression analysis is done in precisely the same way. The only things that change is the source of the data. The regression analyses are run with

the full set of 24 vignettes (now called "cases" in the language of "regression" analysis).

4. The foregoing regression equation in #2 is first estimated for the total panel, then for each self-defined subgroups, and finally for each respondent. As noted above, the 24 vignettes evaluated by each respondent differ from respondent to respondent, but each set of 24 vignettes is structured to be straightforwardly analyzed by OLS regression.

Table 6.3 shows the results for the total panel, and genders.

6.4 Moving Beyond the Individual and the Self-Defined Group to Mind-Sets

The history of consumer research is filled with studies of preferences, and the recognition again and again of the strong differences between people in preferences for simple things like food, for more complex experiences like vacations, and even for the ethereal, such as concepts about moral rights and wrongs (Zimmerman and Croker 2014). Listing these differences would require an encyclopedia. What is important to note is that underneath every calm surface are different currents, different points of view, roiling against each other, even though the surface may seem calm and of unitary construction.

Table 6.3 Data from 101 respondents, showing the coefficients for each element, for Total Panel, and for genders

	Total	Male	Female
Base size	101	57	44
A1	22	20	25
A2	21	17	27
A3	22	19	25
A4	18	17	20
B1	21	21	21
B2	21	21	20
B3	21	22	19
B4	20	22	19
C1	21	18	24
C2	23	22	25
C3	25	25	25
C4	23	21	25
D1	21	21	20
D2	17	19	15
D3	20	20	19
D4	21	23	17

A hallmark of Mind Genomics is the search for groups of respondents who are defined as being similar to each other, not by the pattern of who they say they are, nor the pattern of their behaviors (e.g., what they say they buy), but rather by the way they react. Mind Genomics does not look at variability among people as something noisome, something to be averaged out by running many thousands of respondents for a study, or something to be silenced by removing extraneous sources of variability, to make everyone alike, somehow, except for their "intrinsic" inside. Rather, Mind Genomics seizes upon this variation to offer a way to make sense, to reduce this variation to groups of like-minded individuals, groups which, however, differ from each other. In other words, Mind Genomics welcomes this interindividual variation as a representation of new worlds of the mind.

The underlying principles of Mind Genomics comes from nature and mathematics, as well as from human psychology. Nature tends to be complicated; it is people who want to make this complication simple, this complex plain. Or in the words of composer Georg Friedrich Handel in the Oratorio, "The Messiah," the "rough places plain." It is mathematics that allows the simplification, doing so in elegance, simplicity, and ultimately in effectiveness.

The previous presentation of the data from the different subgroups revealed that there were some elements, which performed strongly. Had we had no idea of the "meaning" of the elements, we might have been encouraged to weave a narrative of findings around the nature of these positive coefficients, such as their magnitudes, the number in each subgroup, and so forth. That is, like so much research where the test stimuli have no "cognitive richness," no "intrinsic meaning," we as researchers would have been left with the task of trying to figure out the "story." Indeed, today's business press of market researchers and sensory professionals again and again talks about telling "stories" from the data, an ability increasing important today in a world where MBA level managers have little patience, and short attention spans.

We now proceed to mind-set segmentation, where the stories emerge, almost with the force of "ocular trauma," namely, hits the researcher "between the eyes." Here are the "facts on the ground."

1. Daily life shows us again and again that people differ from each other. The variability is part of the obvious. What is not so obvious is what to do with this noise, this variability, this random, hard-to-explain departure from what otherwise would be a simple, understandable, perhaps measurable truth. But in the end variability among people exists. Modern psychology, sociology, marketing, and indeed many disciplines dealing with people and behavior have come to realize that people differ from each other in the way they think (Barreto et al. 2021; Smith and Bond 2019).
2. The history of consumer research is filled with studies of preferences, and the recognition again and again of the strong differences between people in preferences for simple things like food, for more complex experiences like vacations, and even for the ethereal, such as concepts about moral rights and wrongs. Listing these differences would require an encyclopedia. What is important to note is that underneath every calm surface are different currents, different points

of view, roiling against each other, even though the surface may seem calm and of unitary construction.

3. A key organizing principle of Mind Genomics is that for every topic there is usually a small set of different clusters, mind-sets, comprising individuals who think differently about the same topic. That is, Mind Genomics does not look at this variability among people as something noisome, something to be averaged out by running many thousands of respondents for a study, or something to be silenced by removing extraneous sources of variability, to make everyone alike, somehow, except for their "intrinsic" inside. Rather, Mind Genomics seizes upon this variation to offer a way to make sense, to reduce this variation to groups of like-minded individuals, groups which, however, differ from each other. In other words, Mind Genomics welcomes this interindividual variation as the reality, as "facts on the ground," suggesting different worlds of the mind.

The tool for creating mind-sets in the world of Mind Genomics is known generically as "cluster analysis" (Mucherino et al. 2009; Kodinariya and Prashant 2013). The wonderful thing about the clustering is that with this tool based only on mathematics, we end up being able to see clear new and meaningful patterns emerge. When we look at the data before it is clustered to find natural groups, we end up with data where some ideas score strongly for one group but score weakly for another group. There "seems to be something," but that "something" is elusive, hard to explain. With clustering, often the story emerges very clearly. Or in the words of Harvard psychologist William James (Goodman 2022), what had been a "blooming, buzzing confusion' when we look at the data from the total panel resolves itself into new, clear patterns easy to understand, easy to label, and easy to use.

The mathematics for clustering is straightforward. One should visualize an Excel matrix, such as that shown in Table 6.4. The columns correspond to the different elements. The rows correspond to the different respondents. The body of this matrix, this spreadsheet, has numbers. The numbers are the coefficients from the individual-level modeling, one set of 16 coefficients for each respondent, that set of

Table 6.4 Coefficients for the first 10 respondents

Res	A1	A2	A3	A4	B1	B2	B3	B4	C1	C2	C3	C4	D1	D2	D3	D4
1	36	31	38	47	24	25	24	26	20	22	28	47	29	11	32	40
2	3	10	22	24	51	55	44	47	21	43	1	21	16	47	36	10
3	23	−18	3	38	−18	74	18	−31	27	10	11	−6	9	14	24	51
4	−1	62	10	−14	−19	−67	−16	2	5	57	8	21	67	67	59	88
5	−1	−71	18	−48	26	−31	34	−12	69	9	111	−11	78	37	33	32
6	27	39	17	34	35	26	36	59	−9	−3	44	25	23	−15	32	53
7	40	35	33	34	38	36	39	28	5	35	20	16	23	38	46	16
8	22	29	49	37	37	29	46	56	30	9	34	34	−2	−12	23	39
9	25	22	25	26	25	43	12	14	43	48	29	20	33	44	40	29
10	54	23	8	44	21	5	2	7	39	38	−26	17	44	22	47	46

coefficients created by using OLS regression (Zdaniuk 2014) to estimate the linear equation for the individual respondent. DV = k1A1 + k2A2 … k16D4. Keep in mind that it is completely correct to work with the 24 cases or vignettes evaluated by each respondent because the construction of the 24 vignettes for each respondent was done in accordance with a permutation of the experimental design. Table 6.4 shows part of the table of coefficients, estimated at the level of each respondent, by OLS regression.

The statistical procedure known as cluster analysis can be applied in an objective manner to this data set, putting each of the 101 respondents into a one group of a pre-set number of groups. The easiest cluster is one big cluster, called the total panel. It is no problem to put all 101 rows into the analysis. Each of the items (viz., people) is assigned to one group only.

Afterward, the clustering program will try to divide the full set of 101 respondent into two, and then three groups. These groups are mutually exclusive; based upon the pattern of coefficients, the cluster program will put the specific respondent into only one of the different clusters.

The clustering program is told how to compute the "dissimilarity" between pairs of "items," viz., people, based on the measures it is given. For our example, the measure is the quantity (1-Pearson R, viz., 1 – Pearson Correlation Coefficient), computed across the 16 corresponding coefficients for each pair of the 101 respondents. The clustering program is then told the number of clusters to discover, but not told anything else about the meaning of the data. All that the clustering program must use is the correlation or strength of linear relation between all pairs of respondents, and the number of clusters or groups to use. The rules are simple:

1. The people or items in a cluster (viz., group) are to be like each other "similar to each other." Operationally, this means that the Pearson Correlation Coefficients, R, are high when computed for people assigned to a cluster.
2. The center points of the different groups or clusters are as different as possible.

Table 6.5 shows the key statistics for the clustering. The first column shows the 16 elements. The second column (F ratio) shows the degree to which the variable is able to separate the elements into clusters (Kodinariya and Prashant 2013). The F ratio can be considered as a signal/noise ratio. The higher the F ratio the stronger the element separates the two clusters. The third and fourth columns show the coefficients for each cluster, computed by the clustering program from the average of coefficient values for all respondents assigned to the cluster. Later on, when it is time for the actual estimation of the coefficients for each cluster, the researcher will not average the individual coefficients from separate equations, one per respondent in the cluster. Rather, the researcher will put all of the data from the respondents of a specific cluster into one OLS regression. The differences between the two approaches are noticeable, but small and generally immaterial.

Table 6.5 Key information about the performance of the 16 elements by cluster

Element	F ratio ability to separate clusters	Average coefficient cluster 1 N = 50	Average coefficient cluster 2 N = 51
A1	4.2	28	17
A2	0.3	26	23
A3	5.2	30	17
A4	2.8	25	16
B1	1.5	12	20
B2	0.0	19	19
B3	7.1	12	28
B4	1.2	20	13
C1	11.2	32	13
C2	13.2	37	18
C3	9.2	33	18
C4	21.4	35	10
D1	28.8	7	38
D2	44.8	−1	34
D3	27.7	7	35
D4	3.3	15	26

6.5 An Actual Study: Setting Up the Mind Genomics Program on the BimiLeap.com Platform

The remainder of this chapter presents a study in detail, showing what the researcher does, from start to finish. We will now revisit the data from the introductory part of this chapter, but disclose the groups in detail, and the results from clustering. We will focus on the process here, talking about what to look for in the results. This section is meant to be "general," instructive, rather than to focus on the results of the experiment. The rest of the book is on results; this section is primarily about "teaching."

The platform, www.BimiLeap.com should be consulted for actual experience with all the steps. The platform, www.BimiLeap, is templated, making the process easier. Templating allows the researcher to do the study, simply by thinking and filling in blanks. Mind Genomics has been augmented with AI, artificial intelligence, to help the researcher develop ideas.

Step 1: Decide on the topic, name the topic, and register the study (Fig. 6.1). It may seem somewhat silly to spend time on naming a study. The reality, however, is that we are not taught to think in school, or at least not taught to think creatively and critically. The act of naming the topic forces the prospective researcher to summarize the idea into a short phrase. Most people summarize the idea by a long sentence and must be taught to condense the thinking into a pair or triple of words, which are telegraphic. It is interesting to watch novice users of Mind Genomics struggle with

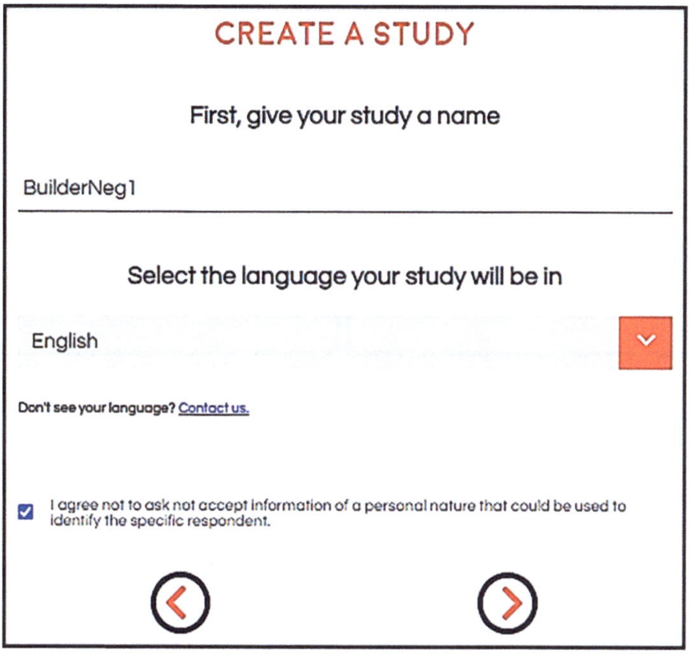

Fig. 6.1 Screenshot showing the first step, name the study. The study name is BuilderNeg1

this issue of naming the study. Fig. 6.1 shows an example of the first step, with the name simply BuilderNeg1, viz., builder negotiation study 1.

Step 2: Develop four questions which "tell a story." Mind Genomics works in a Socratic way. Rather than having the researcher develop an entire study without any aids, Mind Genomics works by the Socratic fashion of question and answer. The researcher must create the four questions which 'tell a story." The task may seem simple, but in reality, this request to "tell a story" appears to frustrate many people, even intimidating some, and contributing to their abandonment of the project. We are simply not taught to ask questions. We certainly answer questions posed to us, but the idea of creating a series of questions is something that must be learned.

The Mind Genomics platform helps a little, by presenting the researcher with the templated request for four questions. Figure 6.2 shows two panels. Panel A shows the space for the questions. Panel B shows the set of four questions.

At this point in the process, many people, professionals, and amateurs, simply give up. The reason is simple. We are not taught to organize our thoughts for new topics. We can organize our thoughts and ask meaningful questions for topics in which we are experts or at least experienced, but the notion of asking questions about a new topic frightens adults. It does not necessarily frighten children, but it frightens professionals and amateurs alike, who worry that they are perhaps asking the "wrong" question or not phrasing the "right question" correctly.

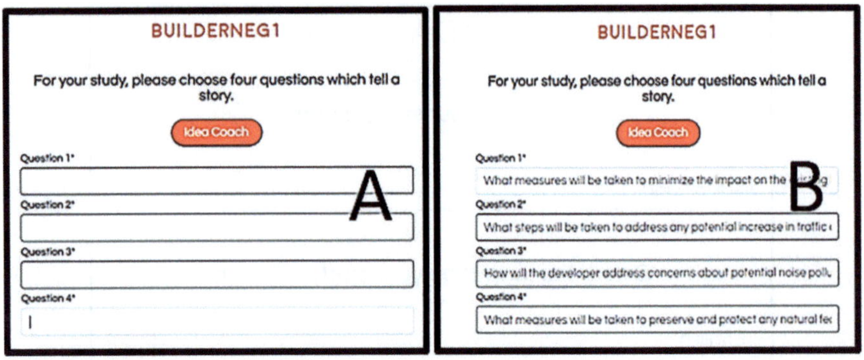

Fig. 6.2 The template for creating four questions which "tell a story"

Step 3: Use artificial intelligence (AI) to suggest questions. As just remarked above, the specter of having to come up with four questions can be daunting. Indeed, during the two-decades long history of Mind Genomics, developing questions became the stumbling block. Researchers were frightened off, often saying they wished there was some way to "jump start" the questions. Of course, those researchers who did the task, coming up with questions, soon found that the task was easier than they had feared. However, the fear of organizing one's ideas and having others somehow "judge' their efforts was frightening.

Going back to about 2010, the original notion was to create a standardized list of questions akin to the types of general questions that a reporter might use to investigate a topic and write a story. This seemingly easy-to-implement system, a "who, what, where, why, when, how" system, proved to be impractical. Although it was straightforward to ask the sequence of questions, the researchers to whom it was presented still felt stymied and expressed their need for a solution that would be more substantive.

The introduction of AI to the world through ChatGPT (OpenAI 2023) boosted Mind Genomics, especially when incorporated into the platform in a way requiring little thinking. When instructed to create four questions, the researcher could type in one or several questions (Fig. 6.2, Panel A and B) or move to a separate screen where the researcher could type something about the topic (Fig. 6.3, Panel A). What was to be typed was left up to the researcher. The objective was to request the embedded AI to present questions about the topic in detail, making the answers short (e.g., 15 words or fewer) and understandable to a young person. The AI would return with sets of 15 questions (Fig. 6.3, Panel B). The researcher could select one or several questions, embed them immediately into the study, edit the questions "on the fly" immediately, or request an additional 15 questions.

Figure 6.3 shows the use of and result of AI through Idea Coach. Panel A provides a space where the researcher can write in the topic description and the request for questions. Panel B shows four of the first 15 questions returned by AI through Idea Coach and the options to select, re-run, or edit the request and rerun.

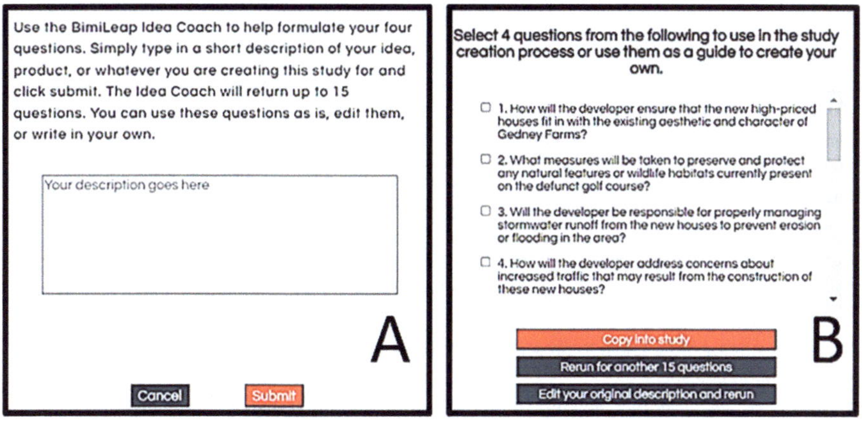

Fig. 6.3 The use of AI and result of AI emerging from Idea Coach

The AI embedded in the Mind Genomics platform, BimiLeap, moves beyond the one-step request and selection of questions. One of the goals of Mind Genomics is to educate the researcher. Recently (spring, 2023), the single step of using AI to suggest questions (Fig. 6.3b) was replaced by creating an "Idea Book."

The Idea Book is set up in Excel® format. Each page presents the effort for a particular iteration. Table 6.6 shows the AI summarization of Idea Coach's first iteration, with the top showing the input to AI, and the bottom showing the detailed information emerging from the AI analysis. Each iteration receives this detailed treatment. The AI has "no memory," treating each iteration as totally new. It is the task of the researcher to guide the AI.

Step 4: For each question generate four answers. The researcher now moves to the next screens. Each screen requires the researcher to provide four answers to each question. The researcher may provide her or his answers, or once again use the Idea Coach to generate answers. It is the job of the researcher to edit the questions, and by doing so create answers, which will be stand-alone phrases painting a word picture.

Note that for some situations, although not the one here, the prompt may be elaborated to begin with the added phrase "Explain in detail," and again concludes with the request to make the answer short (less than 15 words), and understandable to a 12-year-old.

The Idea Book created by Idea Coach (https://www.bimileap.com/home) once again records the 15 suggested answers to each question, for each iteration. Thus, when the researcher user Idea Coach repeatedly, e.g., five times (called iterations) to suggest answers for Question 1, the Idea Book would generate one complete analysis for each iteration, viz., five separate analyses. Each analysis appears in a totally separate tab. The formatted analysis for each iteration would be similar to that appearing in Table 6.6, and thus might in itself provide an "education" in the topic.

Table 6.6 Dialog with AI, showing the input to the AI in bold italics, and the output from AI

Input to AI: Results 1
Topic: I live in White Plains, Gedney Farms, an area with many private houses, and a defunct golf course. The golf course is being sold to a land developer who promises to build single-family, high-priced houses. How should the community interact with the developer to maintain the local environment? Specifics are critical here.
Output from AI
Topic Questions
Can you provide specific details on the design and layout of the proposed single-family houses?
How will the new development integrate into the existing neighborhood in terms of architecture and aesthetics?
What measures will be taken to preserve and protect the existing natural elements, such as trees, wildlife, and open spaces?
Will there be any additional amenities, such as parks, walking trails, or community spaces, included in the development?
How will the new development impact traffic flow and congestion in the area? Will there be any plans for road improvements or traffic management?
What steps will be taken to address any potential noise or pollution issues that may arise from the new development?
Will there be any preservation or restoration of historic structures or landmarks on the property?
How will the new development contribute to the local economy and job market?
What plans are in place to ensure the new houses are energy-efficient and environmentally friendly?
Will there be any efforts to encourage sustainable transportation options within the development, such as bike lanes or public transit access?
How will the developer address concerns about increased demand on local public services, such as schools, healthcare facilities, and public utilities?
What are the details of the developer's timeline for the project, including expected construction phases and completion dates?
Will there be any affordable housing options or contributions to affordable housing initiatives within the development?
How will the developer engage with the community throughout the construction process, addressing concerns and providing updates?
What plans are in place to ensure a smooth transition and integration of the new development into the existing neighborhood?
Key Ideas
Design and layout of the proposed single-family houses
Integration of the new development into the existing neighborhood in terms of architecture and aesthetics
Preservation and protection of existing natural elements
Additional amenities such as parks, walking trails, or community spaces
Impact on traffic flow and congestion, road improvements, and traffic management
Measures to address potential noise or pollution issues
Preservation or restoration of historic structures or landmarks
Contribution to the local economy and job market

(continued)

Table 6.6 (continued)

Energy efficiency and environmental friendliness of the new houses
Encouragement of sustainable transportation options
Addressing increased demand on local public services
Timeline for the project and expected completion dates
Affordable housing options or contributions to affordable housing initiatives
Community engagement throughout the construction process
Smooth transition and integration of the new development into the existing neighborhood.
Themes
1. Design and Integration:
Design and layout of houses
Integration into existing neighborhood
Architecture and aesthetics
Preservation of natural elements
Preservation/restoration of historic structures or landmarks
2. Amenities and Infrastructure:
Additional amenities (e.g., parks, trails, community spaces)
Road improvements and traffic management
Addressing noise and pollution issues
Sustainable transportation options
Increased demand on public services
3. Economic Impact:
Contribution to local economy and job market
Affordable housing options or contributions
Community engagement and smooth transition
4. Energy Efficiency and Environmental Friendliness:
Energy efficiency of new houses
Environmental friendliness of new houses
5. Project Timeline:
Timeline for the project and completion dates
Perspectives
1. Design and Integration:
Plus: Well-designed and visually appealing houses can enhance the overall aesthetic appeal of the neighborhood.
Minus: Integrating new houses into an existing neighborhood can be challenging, as it may disrupt the existing character and architecture.
Interesting: Balancing the preservation of natural elements and historic structures with the need for new development can create unique and innovative design opportunities.
2. Amenities and Infrastructure:
Plus: Additional amenities and community spaces can improve the quality of life for residents and create a sense of belonging.
Minus: Road improvements and increased traffic can temporarily disrupt daily routines and create inconvenience for residents.
Interesting: Sustainable transportation options, such as bike lanes or electric vehicle charging stations, can promote environmentally friendly commuting.

(continued)

Table 6.6 (continued)

3. Economic Impact:
Plus: New construction projects often create job opportunities and boost the local economy.
Minus: Increased demand for public services, such as schools and healthcare facilities, may strain existing resources.
Interesting: Involving the community in decision-making processes and ensuring a smooth transition can foster a strong sense of community pride and ownership.
4. Energy Efficiency and Environmental Friendliness:
Plus: Energy-efficient houses can reduce utility costs for residents and contribute to a more sustainable future.
Minus: Implementing energy-efficient features may involve additional upfront costs for developers and potential challenges in persuading buyers to pay higher prices for eco-friendly homes.
Interesting: Exploring innovative technologies and materials to create environmentally friendly houses can lead to groundbreaking discoveries and industry advancements.
5. Project Timeline:
Plus: Clear timelines for the project can provide certainty and facilitate effective planning for all stakeholders involved.
Minus: Delays in construction or unexpected setbacks can cause frustration and inconvenience for residents.
Interesting: The project timeline may vary depending on various factors, such as weather conditions, availability of resources, and coordination with local authorities. Managing these factors effectively can showcase effective project management skills.
What Is Missing
What is the expected price range for the new single family houses?
What is the size and square footage of the proposed houses?
What is the total number of houses or units planned for the development?
Will there be any homeowner association fees or community dues associated with the new development?
Will there be any restrictions or guidelines on home design or architectural style for the new houses?
How will the new development impact property values in the surrounding area?
Are there any plans for retail or commercial spaces within the development?
Will there be any public or private schools included in the development plans?
What measures will be taken to ensure the safety and security of the new development and its residents?
Are there any plans for public art or cultural elements within the development?
What is the anticipated timeline for the completion of any proposed amenities or community spaces?
Will there be any public meetings or forums for community input and feedback on the development plans?
What are the plans for landscaping and green spaces within the development?
Will there be any stormwater management or drainage systems incorporated into the development plans?
What impact will the new development have on the existing property values and taxes in the area?

(continued)

Table 6.6 (continued)

Will there be any efforts made to incorporate sustainable building materials or construction methods in the development?
What are the proposed plans for street lighting and neighborhood safety features?
Will there be any regulations or guidelines in place regarding noise levels or construction hours during the development process?
What will the impact of the new development be on the surrounding infrastructure, such as water and sewer systems?
Will there be any environmentally friendly features, such as solar panels or rainwater harvesting, included in the new houses?
Alternative Viewpoints
What are the potential negative effects of the new development on property values in the existing neighborhood?
Are there concerns about the potential overcrowding or strain on local infrastructure?
What are the potential environmental impacts of the new development, such as increased water usage or habitat disruption?
Are there any potential health or safety concerns associated with the new development, such as increased noise or air pollution?
What are the potential impacts of the new development on nearby businesses or local commerce?
Are there concerns about the loss of privacy or changes to the overall character of the existing neighborhood?
What mitigation measures are in place to address any potential flooding or drainage issues caused by the new development?
Are there any concerns about the increased demand for public transportation or parking in the area?
What are the potential economic benefits or drawbacks of the new development on the surrounding property values and local businesses?
What measures will be taken to address any potential increase in crime or security risks associated with the new development?
What are the potential effects of the new development on the existing schools, in terms of increased enrollments or strain on resources?
Are there concerns about the impact of the new development on the natural beauty or scenic views of the area?
What are the potential impacts of the new development on the local wildlife and ecosystem?
Are there any efforts to involve local businesses or contractors in the construction process to stimulate the local economy?
What are the potential long-term effects of the new development on the surrounding natural resources, such as water supply or air quality?
Are there concerns about the potential loss of community character or sense of identity in the existing neighborhood?
What are the potential social or cultural impacts of the new development, in terms of demographic changes or changes to the community fabric?
What are the potential impacts of the new development on the existing traffic patterns and transportation infrastructure?
Are there concerns about the potential increase in property taxes or other financial burdens associated with the new development?

(continued)

Table 6.6 (continued)

Interested Audiences
1. Potential home buyers: They would be interested in the specific details of the design and layout of the houses to determine if it meets their preferences and needs.
2. Current residents of the neighborhood: They would be concerned about how the new development will fit in with their existing homes and if it will maintain the overall aesthetic of the area.
3. Environmentalists and conservationists: They would be interested in knowing how the natural elements in the area will be preserved and protected during the construction process.
4. Families with children: They would be concerned about the availability of amenities such as parks, walking trails, and community spaces for recreation and socialization.
5. Local commuters: They would want to know how the new development will impact traffic flow in the area and if any measures will be taken to improve road conditions.
6. Health advocates: They would be concerned about any potential noise or pollution issues that may arise from the construction and want to know how they will be addressed.
7. Historians and preservationists: They would be interested in any plans to preserve or restore historic structures or landmarks on the property.
8. Local business owners: They would be interested in how the new development will contribute to the local economy and potentially stimulate job growth.
9. Environmental activists: They would want to know the measures that will be taken to ensure energy efficiency and environmental friendliness of the new houses.
10. Sustainable transportation advocates: They would be interested in any plans to promote sustainable transportation options within the development, such as bike lanes or access to public transit.
11. Local public service providers: They would be concerned about the impact of the new development on public services and want to know if any plans are in place to address increased demand on schools, healthcare facilities, and public utilities.
12. Community members: They would be interested in the developer's timeline for the project, as well as how the developer plans to engage with and address concerns of the community during the construction process.
13. Affordable housing advocates: They would want to know if there will be any affordable housing options or contributions to affordable housing initiatives within the development.
14. Potential investors: They would be interested in knowing the details of the developer's plans and timeline for the project and how it may impact property values in the area.
Opposing Audiences
1. Existing homeowners in the neighborhood may oppose the topic if they are concerned about the potential changes in the neighborhood's character or property values.
2. Environmental activists may oppose the topic if they believe that the development will have a negative impact on the natural elements, such as the destruction of trees or disruption of wildlife habitats.
3. Local businesses may oppose the topic if they fear increased competition or a decrease in their customer base due to the development.
4. Residents who prioritize traffic flow and congestion may oppose the topic if they believe that the new development will worsen these issues in the area.
5. Preservationists and historians may oppose the topic if they believe that the new development will disregard or destroy any existing historic structures or landmarks.
6. Affordable housing advocates may oppose the topic if they believe that the development does not adequately address the need for affordable housing options in the area.

(continued)

Table 6.6 (continued)

Innovations
1. Design and Integration:
Green roofs and living walls for incorporating natural elements into buildings
Adaptive reuse of existing structures to preserve historic landmarks
Use of sustainable building materials and construction techniques
2. Amenities and Infrastructure:
Smart city technology for improved traffic management and infrastructure efficiency
Bike-sharing or electric vehicle charging stations for sustainable transportation options
Implementation of noise-reducing measures such as sound barriers or strategic landscaping
3. Economic Impact:
Incentives for local businesses or entrepreneurs to set up shops or services within the new neighborhood
Job training programs or apprenticeships for local residents to secure employment opportunities created by the project
Creation of a community fund to support affordable housing initiatives or community development projects
4. Energy Efficiency and Environmental Friendliness:
Solar panel installation for renewable energy generation in new houses
Rainwater harvesting systems for landscape irrigation and water conservation
Use of energy-efficient appliances and smart home technology to monitor and reduce energy consumption
5. Project Timeline:
Utilization of project management software or tools for efficient coordination and tracking of project milestones
Implementation of modular construction techniques to accelerate the building process
Collaboration with local universities or research institutions for innovative design or construction methods.

Figure 6.4 shows the first question selected by the researcher. The researcher now must develop four relevant answers to the question. One strategy is for the researchers to provide the four answers to the question. A second strategy is to request that AI in the form or Idea Coach simply work with the question as provided to the BimiLeap program by the researcher (previous phase), and now simply let the AI do all the work. The questions are already "part of the study." It is not difficult to instruct AI to take the question as currently embedded in the study and come up with 15 answers.

Figure 6.4 does exactly that. Panel A (top) shows Question 1 of 4 (What measures will be taken to minimize the impact on the existing local wildlife and ecosystem). Wherever these questions came from does not matter. By using Idea Coach, the researcher automatically requests 15 answers, 6 of which are shown in Panel B. The remaining answers are accessed by scrolling through the screen.

Table 6.7 shows the four questions and four answers to each question after the researcher has edited and polished them using the AI suggestions. It is relevant to note here that the approach presented by Mind Genomics and the incorporation of

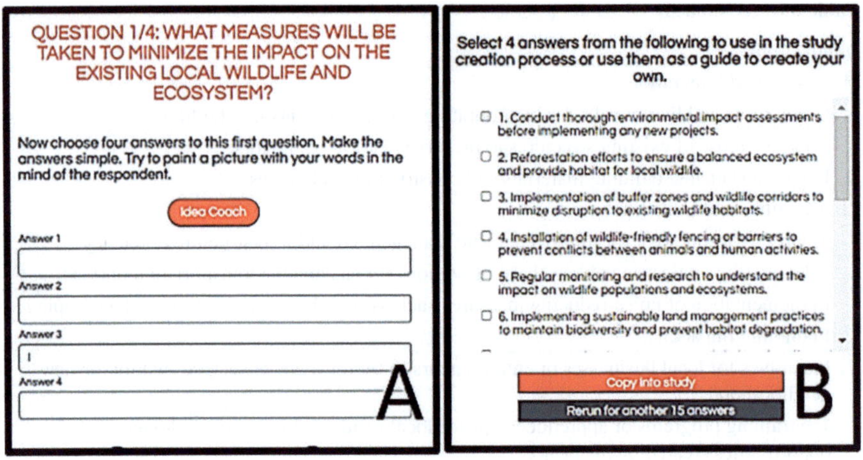

Fig. 6.4 Panel A. The request to provide four answers to question 1 and the output from AI

AI through Mind Coach provides a different way to "do science." We are accustomed to research often done to answer specific questions emerging from previous studies. The effort is part of the so-called hypothetico-deductive system (Lawson 2000), where science advances little steps at a time. The Mind Genomics effort is somewhat different, focusing on granular issues to be sure but expanding those granular issues to cover a wider net of attitudes and behaviors. Toward that end, the questions and answers from the AI, viz., Idea Coach, teach the topic, not just a set of questions.

Step 5: Develop self-profiling classification questions. Beyond the "experiment" to understand the respondent's mind is the issue of WHO the respondent is, what does the respondent DO, and how does the respondent THINK? We are not talking here about the experiment itself but rather about the information provided by the respondent in a set of questions. These questions will be used to create self-defined subgroups of respondents. The researcher will be able to understand the responses to the vignettes from each group separately. Table 6.8 shows these questions created by the researcher. Questions about age and gender are automatically included. The questions themselves will appear as a pulldown menu at the start of the study.

Figure 6.5 shows the two panels dealing with the self-profiling classification. Panel A on the left shows the first question (large rectangle) and the four answers provided by the user (four smaller rectangles). The user can select from 2–8 questions, and each question can have 2–8 answers. Panel B shows the self-profiling classification as it appears to the respondent in the form of a pull-down menu. Once the question is selected, the screen changes to show the options. The respondent must select one of the options.

Step 6: Create the introduction to the study. Most respondents have yet to participate in a Mind Genomics study at the time of this writing, Spring, 2024. Consequently, when the respondent is recruited to participate, it is essential to give the respondent

Table 6.7 The four questions and four answers to each question after the researcher has edited and polished them

	Question A: What measures will be taken to minimize the impact on the existing local wildlife and ecosystem?
A1	Implement habitat restoration and enhancement programs to support local wildlife populations.
A2	Encourage responsible waste management practices to prevent pollution of local ecosystems.
A3	Introduce measures to minimize vehicle collisions with wildlife, such as wildlife crossings and signage.
A4	Establish buffer zones around construction sites to prevent disturbances to wildlife.
	Question B: What steps will be taken to address any potential increase in traffic or congestion that the new development may cause?
B1	Collaborate with local transportation agencies and traffic management departments to analyze traffic patterns and identify potential bottlenecks.
B2	Implement smart traffic signal systems that can adjust timings based on congestion levels to optimize traffic flow.
B3	Develop a comprehensive transportation demand management plan that encourages carpooling, ridesharing, and the use of electric vehicles.
B4	Promote flexible working hours to stagger peak travel times and reduce congestion during rush hour.
	Question C: How will the developer address concerns about potential noise pollution during construction and after the houses are built?
C1	Implement physical barriers or soundproofing measures, such as temporary walls or fences, to minimize noise pollution during construction.
C2	Provide a dedicated hotline or communication channel for residents to report noise-related issues, ensuring a swift response to address concerns.
C3	Orient houses away from potential noise sources, such as major roads or industrial areas, to minimize noise impact on residents.
C4	Establish a community noise complaint resolution process, where residents can voice their concerns, and ensure that their issues are addressed promptly and effectively.
	Question D: What measures will be taken to preserve and protect any natural features or wildlife habitats on the golf course during and after development?
D1	Create nesting areas, bird boxes, and other structures to encourage wildlife to thrive within the golf course environment.
D2	Plant native vegetation and trees to enhance natural habitats and provide food and shelter for wildlife.
D3	Establish designated wildlife corridors throughout the golf course to maintain connectivity between habitats.
D4	Install bird-friendly designs in golf course structures, such as windows with minimal reflection or visual cues to minimize bird collisions.

some instructions about what to do but only a little information beyond that. The objective is to have the respondent evaluate the test stimulus using a simple scale. The actual test stimuli comprise a set of 24 different vignettes, a vignette being a combination of elements, with a minimum or two elements in a vignette, and a maximum of four elements. The actual construction of the vignette is described

Table 6.8 The self-profiling classification questionnaire, developed by the researcher to understand the respondent

What is the type of home in which you live?
1 = Apartment 2 = Condo 3 = Attached row house 4 = Private house
Where do you live?
1 = Big city 2 = Small city 3 = Suburb 4 = Rural
Who owns the place where you live?
1 = Landlord 2 = Cooperative board 3=Condominium - I own my condo 4=I own my own home - paying mortgage 5=I own my own home - paid off mortgage
How do you feel about housing construction?
1 = Not concerned 2=Bothered by the inconvenience during construction 3 = Bothered by the effect on the land 4=Bothered by potential crowding
Describe which phrase fits you best?
1 = I am not particularly involved in my community affairs
2 = I know what's going on in the community but that's about it
3 = I am active in community affairs

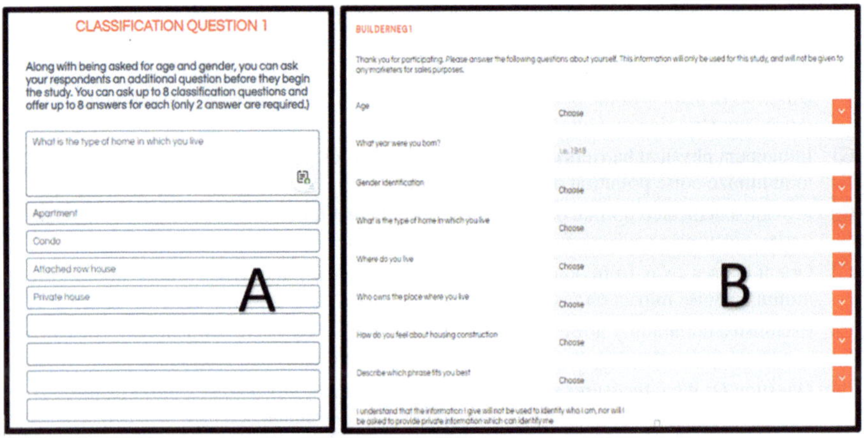

Fig. 6.5 An example of one question (Panel B) and the pulldown menu (Panel B)

below. For now, it is sufficient to give the respondent simple instructions, requesting that the respondent read the vignette and assign a rating using the scale provided.

Most respondents require very little information about the ultimate purpose of the study. Sometimes, the researcher feels compelled to introduce the study in depth. Short introductions to the study are just as effective as long-winded introductions. Respondents want to know what to do when completing the study. Why the study is being run, the ultimate use of the data and the moral benefits of participating in the study are quite irrelevant and might even be biased. The less said, the better, although sometimes the respondent needs to know a lot, such as the background to Mind Genomics used in a law or business-legal case.

Figure 6.6 presents the orientation given to the respondent. Figure 6.8 presents more information than is usually the case because we are dealing with a complex, almost a business-legal issue.

Step 7: Create a rating scale for the evaluation of test vignettes. The essential action in the Mind Genomics study is the evaluation of 24 vignettes, defined as a combination of raw materials, the elements shown in Table 6.7.

Table 6.9 shows the questions that are used. Figure 6.7 shows a screenshot of the page where the researcher fills in a short rating question and then the scale values for the five rating points. As we see in Table 6.9, the scale considers two types of responses: the nature of the builder concession (good vs. poor) and the nature of the community response (accept vs. reject). This "two-sided" rating scale allows the researcher to capture more of the dynamics of the different offers, creating a richer database.

Step 7: Create an open-ended question, to appear after the evaluation of the vignettes. The researcher has the opportunity to have the respondent answer an open-ended question about any aspect of the topic or anything else. This information will be collected and put into the database but not further analyzed by the BimiLeap platform. The responses can be analyzed by other text and analyzers and summarizers, such as QuillBot (Fitria 2021). Figure 6.8 shows the screenshot for the open-ended question.

Step 8: Select the source of respondents. The BimiLeap platform has several ways to acquire respondents to participate. These different options are shown in Fig. 6.9. The researcher selects the source of respondents and, where appropriate, specifies the nature of the respondents. BimiLeap.com provides a set of qualifications across different geo-demographic variables. It is left to the researcher to specify the nature of WHO will participate, whether the WHO comes from a commercial panel provider or the respondents available separately to the researcher.

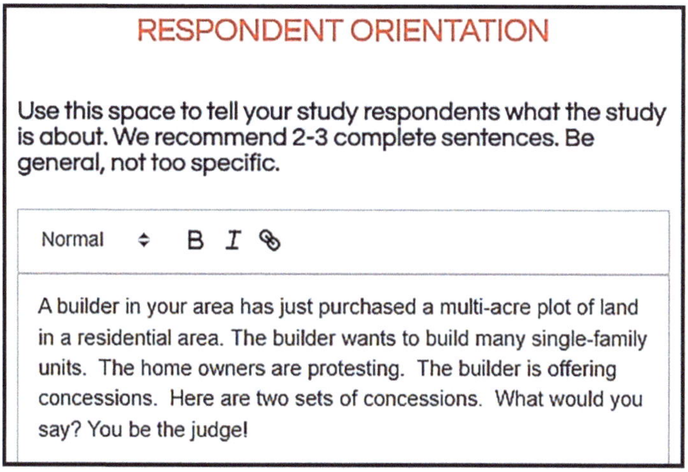

Fig. 6.6 The respondent orientation

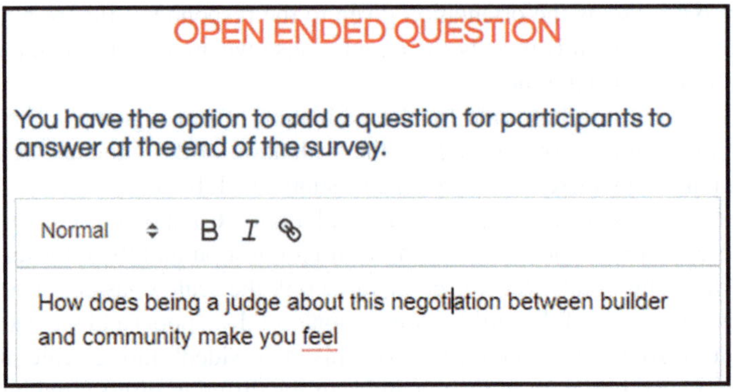

Fig. 6.8 Screenshot for the open-ended question

Table 6.9 The scale points for the 5-point rating scale

1 = Poor builder concessions AND community rejects
2 = Poor builder concessions BUT community accepts
3 = I don't know
4 = Good builder concessions BUT community rejects
5 = Good builder concessions AND community accepts

Step 9: The vignette. The final step to set up the study creates the vignettes. Figure 6.10 shows an example of the vignette. The top of Fig. 6.10 shows the rating question and the scale. The bottom of Fig. 6.10 shows the 2–4 elements put together according to an underlying experimental design. There is no effort to link the three elements into a coherent whole.

The experimental design prescribes the precise construction of the vignette, with most vignettes comprising four elements, some comprising three, and some comprising two. Recall that in the language of Mind Genomics, the word "element" is identical to "answer" or "message." Finally, the elements of the vignette are "standalone" phrases.

To many professionals, the spare appearance of the vignettes is disturbing. There are no connectives between the elements, irritating many professionals who want to see a polished vignette. There is a reason for this. The spare look makes it easier for the respondent to find information and decide. The respondent can "graze" and pick up information rather than having to read the vignette. Were there to be connectives linking the sentences, the respondent would be tired simply from "hacking through the weeds," the verbiage. The goal is not the verbiage but the combination of ideas. Exit interviews with respondents over the years reveal little in the way of complaints

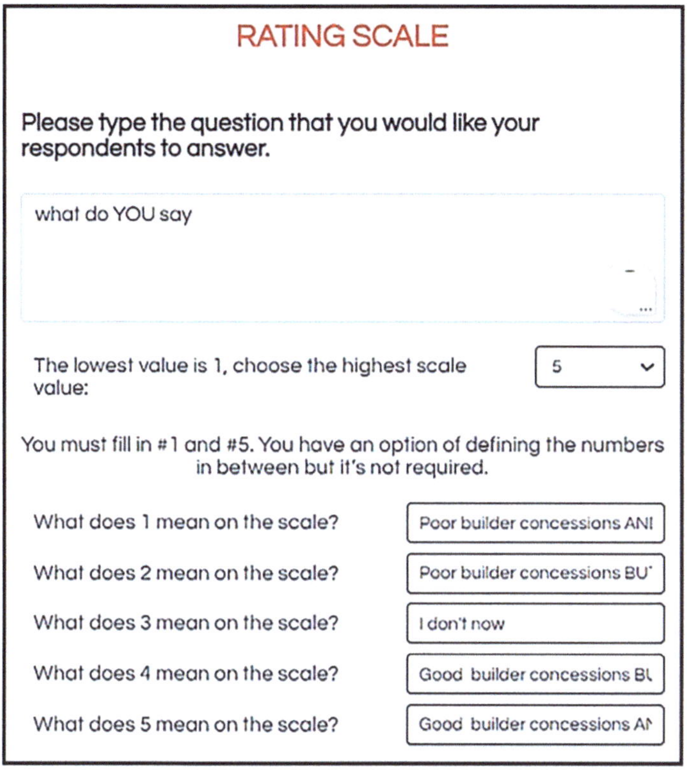

Fig. 6.7 Screenshot of the 5-point rating scale, with the places for the short rating question, and the five scale points

about "fatigue." However, many complain about not being able to discern the "right answer," leading respondents to state that they were simply "guessing" their way through the session.

Step 10: Field the study with respondents. Each respondent receives an invitation to participate, whether the invitation comes from a company specializing in recruiting responses for the so-called "on-line surveys," or whether the invitation comes from colleagues and friends, or even from downloading a link on social media. Experience suggests that it is far less frustrating to work with a company that sources the respondents at what ends up being a low price and fast turnaround.

The data that emerges comprises 24 rows of data for each respondent. Each row is devoted to a specific vignette evaluated by the respondent. The information available is the respondent-level information from the self-profiling classification (see Fig. 6.5), along with the structure of the vignette (see Table 6.1), and finally the rating (see Table 6.9).

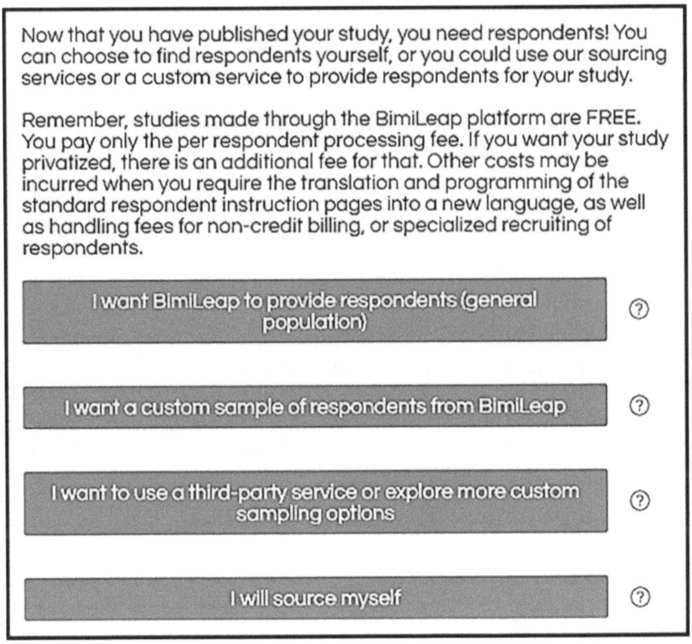

Fig. 6.9 How the researcher "sources" respondents

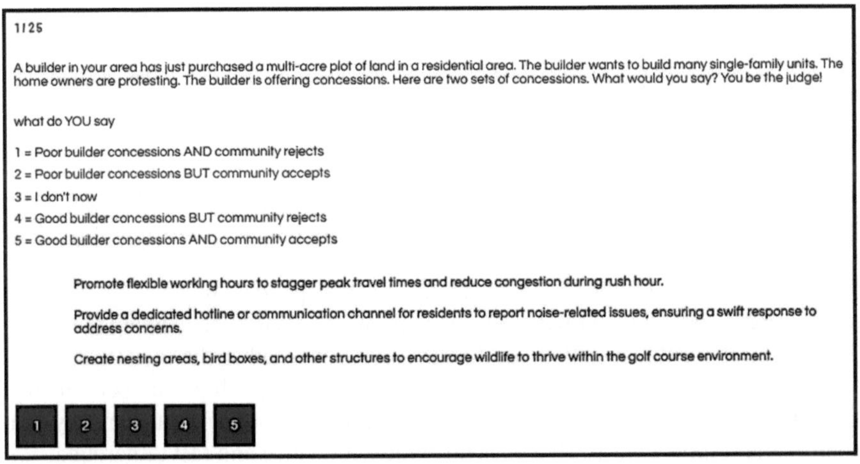

Fig. 6.10 Screenshot of a three-element vignette shown to a respondent. The top element shows the rating scale, and the bottom of the vignette shows the three elements and the rating scale

Step 11: Create new binary-dependent variables and run OLS regression to create the equation. It is important to transform the rating scale into a format that can be analyzed to create meaningful results. The transformation produces two scales.

The objective of Step 10 is to discover how the 16 elements "drive" the dependent variable. The dependent variable is generally a newly created variable that takes on two values, 0 or 100, respectively. Table 6.10 shows the different transformations that can be made in order to create various dependent variables. The value of these different transformations is that one scale can generate a variety of different descriptions of behavior. That capability means that in a short period, the 3–4 min of interaction with the respondent, the researcher can obtain different types of information.

Statistical Note: A vanishingly small random number is added to each newly created dependent variable (<10–5). The rationale for this prophylactic step is that it ensures minimal variability for each newly created binary variable. That minimal variability ensures that the OLS regression will run without problem.

The Mind Genomics platform, BimiLeap.com, also records response time, the time in seconds between the presentation of the vignette on the respondent's computer screen, and the rating assigned by the respondent. That time is measured to the nearest hundred of a second. Response times of 9 s or longer are truncated to 9 s simply because it makes no sense for the respondent to require 9 s to read the vignette. The assumption is that the respondent is multitasking and not paying attention to the task.

Step 12: Select the appropriate dependent variable from the set of newly created binary variables. The Mind Genomics platform is programmed to use R54X (Top2) and R21X (Bot2) as the dependent variable, but other variables can be created "on the fly" and used.

Step 13: Run the OLS regression, after selecting the binary variable to be the dependent variable. The most critical moment in the preparation has now arrived, viz., estimate the relation between the presence/absence of the 16 elements and the binary-dependent variable. Table 6.11 shows the coefficients when we use allow, the different binary variables and RT (response time) to become the dependent variable. We write the equation as: Dependent variable = k1A1 + k2A2 … k16D4. The coefficients are estimated using OLS, ordinary least-squares regression. The equation is estimated without the additive constant, a practice emerging from the need to compare the coefficients from study to study, from person to person. The specific regression also has the name "dummy variable regression" because the 16 independent

Table 6.10 Transformations often made on the 5-point scale to create new binary variables

Variable name	Ratings transformed to 100	Ratings transformed to 0
R5X good concession and offer accepted	5	1,2,3,4
R54X good concession	5,5	1,2,3
R52X offer accepted	5,2	1,3,4
R3X Can't answer	3	1,2,4,5
R41 offer rejected	4,1	2,3,5
R21X poor concession	2,1	3,4,5
R1 = poor concession and offer rejected	1	2,3,4,5

Table 6.11 Parameters for the total panel relating the presence/absence of the 16 elements to the binary-dependent variables and to response time (RT)
The table is sorted by the coefficients for R54x (Good Concession)

	TOTAL PANEL	**R5X**	**R54X**	**R52X**	**R41X**	**R21X**	**R3X**	**RT**
	Builder Concession	**Good**	**Good**			**Poor**	**Do Not Know**	**Response Time Sec**
	Neighborhood Response	**Accept**		**Accept**	**Reject**			
	Total Panel	**R5X**	**R54X**	**R52X**	**R41X**	**R21X**	**R3X**	**RT**
C3	Orient houses away from potential noise sources, such as major roads or industrial areas, to minimize noise impact on residents.	16	25	17	9	2	5	0.8
C2	Provide a dedicated hotline or communication channel for residents to report noise-related issues, ensuring a swift response to address concerns.	14	23	15	13	4	3	0.7
C4	Establish a community noise complaint resolution process, where residents can voice their concerns, and ensure that their issues are addressed promptly and effectively.	15	23	18	9	4	3	0.8
A1	Implement habitat restoration and enhancement programs to support local wildlife populations.	15	22	17	8	3	3	1.2

variables are "dummy values," either 0 (absent from the vignette) or 1 (present in the vignette). No other information is need about these elements other than they either appear or do not appear in a specific vignette, or "case" in the language of regression analysis.

Statistical note: When an additive constant is estimated that additive constant becomes a complicating factor, affecting the magnitude of the coefficients and, in turn, reducing the comparability of the elements based on the sizes of their coefficients. It is simply easier to estimate the equation without the additive constant, an act also called "fitting a model through the origin." The coefficients estimated without an additive constant versus with an additive constant correlate highly. It is only the magnitude of the coefficients which change.

Further are observations emerging from Table 6.11.

One of the most important things to look for is the strength of the elements. Are any really strong elements having a coefficient of 21 or higher for the newly created binary variables? The operational definition of 21 comes from experience. A coefficient of +12 or higher is statistically "significant" but not necessarily meaningful. Experience suggests that coefficients around 20 or 21 correspond to meaningful results that can be interpreted and make sense. R54X is the only newly created binary variable with consistently high coefficients.

For other dependent variables, we may have to lower the criterion, since we are on a "search" for patterns, rather than attempting to confirm or disconfirm a hypothesis. The Mind Genomics efforts try to discover strong elements so that the underlying pattern, the underlying story, can reveal itself. That pattern may not emerge when we make the criterion very high, viz., the coefficient value of +21 that we used for R54X.

The cutoff points are fluid, left to the researcher, and meant only as rules of thumb to allow patterns to emerge. It is clear that many of the elements lead to the neighborhood committee accepting the concessions. It is perfectly permissible for the researcher to make the criterion higher or lower, such as a coefficient of 18 rather than, say, a coefficient of 15. Once again, the purpose of the shading and the higher criterion is to allow patterns to emerge.

Step 14: Move from the Total Panel to subgroups as respondents define them. The self-profiling classification (Table 6.8) allows the researcher to find out more information about the respondent, information that might not otherwise be obtainable. The current version of Mind Genomics requires that the respondent selects one answer from the set of provided answers to each question. Another possibility, not yet implemented, is "check all that apply" the so-called CATA in the world of consumer and sensory research.

Tables 6.12, 6.13, 6.14 and 6.15 show the results for the binary variable R54, this time with the regression run separately for each of the subgroups defined by the classification questionnaire. Note that the base sizes are not equal. Some subgroups have a larger number of respondents, with the coefficients being relatively stable. Other subgroups may have fewer respondents, so the coefficients are less stable due to the small "base size." *A good practice is to ignore subgroups of size nine or lower, but again, this practice is more a "judgment call"* from experience than a rule that has been developed through deep consideration.

The interpretation of subgroups is often tricky. Often the coefficients fail to produce a pattern which makes sense, which tells a story. There may be a story "hiding" in the data, but there is so much variation introduced by small base sizes that the

Table 6.12 The coefficients for R54X by total and gender. The table is sorted by Total Panel

			Gender	
	Coefficient for R54X (Good concession)	**Total**	**Male**	**Female**
	Base Size	101	57	44
C3	Orient houses away from potential noise sources, such as major roads or industrial areas, to minimize noise impact on residents.	25	25	25
C2	Provide a dedicated hotline or communication channel for residents to report noise-related issues, ensuring a swift response to address concerns.	23	22	25
C4	Establish a community noise complaint resolution process, where residents can voice their concerns, and ensure that their issues are addressed promptly and effectively.	23	21	25
A1	Implement habitat restoration and enhancement programs to support local wildlife populations.	22	20	25
A3	Introduce measures to minimize vehicle collisions with wildlife, such as wildlife crossings and signage.	22	19	25
A2	Encourage responsible waste management practices to prevent pollution of local ecosystems.	21	17	27
B1	Collaborate with local transportation agencies and traffic management departments to analyze traffic patterns and identify potential bottlenecks.	21	21	21
B2	Implement smart traffic signal systems that can adjust timings based on congestion levels to optimize traffic flow.	21	21	20
B3	Develop a comprehensive transportation demand management plan that encourages carpooling, ridesharing, and the use of electric vehicles.	21	22	19
C1	Implement physical barriers or soundproofing measures, such as temporary walls or fences, to minimize noise pollution during construction.	21	18	24
D1	Create nesting areas, bird boxes, and other structures to encourage wildlife to thrive within the golf course environment.	21	21	20
D4	Install bird-friendly designs in golf course structures, such as windows with minimal reflection or visual cues to minimize bird collisions.	21	23	17
B4	Promote flexible working hours to stagger peak travel times and reduce congestion during rush hour.	20	22	19
D3	Establish designated wildlife corridors throughout the golf course to maintain connectivity between habitats.	20	20	19
A4	Establish buffer zones around construction sites to prevent disturbances to wildlife.	18	17	20
D2	Plant native vegetation and trees to enhance natural habitats and provide food and shelter for wildlife.	17	19	15

coefficients randomly vary, often being too low to "notice." In these cases, the researcher focuses on the elements with high coefficients and does not realize that some of these "signals," these high coefficients, are only there because of random variability and not because they are part of a real story. Parenthetically, it is these large swings in the value of the coefficient for Mind Genomics, but also in the values of other dependent variables, which call for large base sizes in order to "average out" the random swings, which in turn lets the "signal" or pattern come through.

Table 6.13 shows the coefficients for R54X by total, by type of home, and by where the respondent lives.

Table 6.13 The coefficients for R54X by total, by type of home, and by where the respondent lives
The table is sorted by Total Panel

	Coefficient for R54X (Good concession)		What is the type of home in which you live			Where do you live		
	Group (Binary Ratings)	Total	Apartment	Attached row house	Private house	Big city	Suburb	Rural
	Base Size	101	29	10	55	57	24	11
C3	Orient houses away from potential noise sources, such as major roads or industrial areas, to minimize noise impact on residents.	25	27	36	23	30	21	17
C2	Provide a dedicated hotline or communication channel for residents to report noise-related issues, ensuring a swift response to address concerns.	23	30	13	22	29	15	19
C4	Establish a community noise complaint resolution process, where residents can voice their concerns, and ensure that their issues are addressed promptly and effectively.	23	27	12	22	24	17	30
A1	Implement habitat restoration and enhancement programs to support local wildlife populations.	22	25	11	24	20	27	26

Table 6.15 shows the coefficients for R54X by total, by feeling about housing construction, and by community involvement. The table is sorted by Total Panel.

Step 11: Discover profound patterns (mind-sets) by clustering the coefficients. The history of consumer research is filled with studies of preferences and the recognition again and again of the substantial differences between people in preferences for simple things like food, for more complex experiences like vacations, and even for the ethereal, such as concepts about moral rights and wrongs. Listing these differences would require an encyclopedia. What is important to note is that underneath every calm surface are different currents and points of view, roiling against each other, even though the surface may seem calm and of unitary construction.

The previous presentation of the data from the different subgroups revealed that many elements performed strongly, although there were no clear patterns. Had we had no idea of the "meaning" of the elements, we might have been encouraged to

Table 6.14 The coefficients for R54X by total, and by the nature of the home ownership
The table is sorted by Total Panel

			Nature of home ownership			
	Coefficient for R54X (Good concession)	**Total**	**Landlord**	**Condominium - I own my condo**	**I own my own home - paying mortgage**	**I own my own home - paid off mortgage**
	Base Size	101	19	10	37	30
C3	Orient houses away from potential noise sources, such as major roads or industrial areas, to minimize noise impact on residents.	25	26	26	19	27
C2	Provide a dedicated hotline or communication channel for residents to report noise-related issues, ensuring a swift response to address concerns.	23	17	27	24	22
C4	Establish a community noise complaint resolution process, where residents can voice their concerns, and ensure that their issues are addressed promptly and effectively.	23	24	27	20	24
A1	Implement habitat restoration and enhancement programs to support local wildlife populations.	22	27	23	25	19

weave a narrative of findings around the nature of these positive coefficients, such as their magnitudes, the number in each subgroup, and other characteristics. That is, like so much research where the test stimuli have no "cognitive richness," no "intrinsic meaning," thus leaving researchers with the onerous task of figuring out the story or, more often than would be admitted, synthesizing a story so as not to disappoint. When one looks at the business press of market researchers and sensory professionals, one again sees the emphasis on the ability to tell "stories" from the data, an ability increasingly important today in a world where MBA-level managers have little patience and short attention spans.

Mind Genomics does not look at variability among people as something noisome, to be averaged out by running many thousands of respondents for a study, or to be silenced by removing extraneous sources of variability. Instead, Mind Genomics seizes upon this variation to offer a way to make sense, to reduce this variation to groups of like-minded individuals who differ from each other. A fundamental organizing principle of Mind Genomics is that there is usually a small set of clusters and mind-sets for every topic, comprising individuals who think differently

Table 6.15 The coefficients for R54X (good builder concession) by total, by feeling about housing construction, and by community involvement. The table is sorted by Total Panel

			How do you feel about housing construction				Describe which phrase fits you best		
	Coefficient for R54X (Good concession)	Total	Not concerned	Bothered by the inconvenience during construction	Bothered by the effect on the land	Bothered by potential crowding	I am not particular involved in my community affairs	I know what's going on in the community but that's about it	I am active in community affairs
	Base Size	101	25	33	25	18	28	34	39
C3	Orient houses away from potential noise sources, such as major roads or industrial areas, to minimize noise impact on residents.	25	17	28	29	27	24	18	32
C2	Provide a dedicated hotline or communication channel for residents to report noise-related issues, ensuring a swift response to address concerns.	23	9	29	23	33	19	27	22
C4	Establish a community noise complaint resolution process, where residents can voice their concerns, and ensure that their issues are addressed promptly and effectively.	23	13	25	23	31	14	29	22
A1	Implement habitat restoration and enhancement programs to support local wildlife populations.	22	33	21	15	21	27	24	17

about the same topic. In other words, Mind Genomics welcomes this interindividual variation as a representation of new worlds of the mind.

We now proceed to mind-set segmentation, where the stories emerge, almost with the force of "ocular trauma," namely, hitting the researcher "between the eyes." The underlying principles of Mind Genomics come from nature, mathematics, and human psychology. Nature tends to be complicated, but people want to make this complication simple, this complex plain. Or, in the words of composer Georg Friedrich Handel in the Oratorio, "The Messiah," the "rough places plain." Mathematics allows simplification, doing so in elegance, simplicity, and, ultimately utterly effective.

The mathematics for clustering is straightforward. One should visualize an Excel matrix. The columns correspond to the 16 different elements. The rows correspond to the different respondents. The body of this matrix, this spreadsheet, has numbers. The numbers are the coefficients from the individual-level modeling, one set of 16 coefficients for each respondent. Thus, this matrix contains one row for each respondent, that row showing the individual coefficients.

The set of coefficients is created using OLS regression to estimate the linear equation for the individual respondent. The equation is expressed as before, viz., in the case of the total panel and the critical self-declared subgroups: DV = k1A1 + k2A2 … k16D4. Table 6.4 above shows an example of a matrix of coefficients for 10 individual respondents and 16 elements.

The initial construction of the 24 vignettes for each respondent was done following a permutation of the experimental design, allowing the researcher to perform the OLS regression analysis at the level of each of the 101 respondents. Finally, the prophylactic addition of the vanishingly small random number ensured that there would be the requisite variability in the dependent variable, namely, the newly created binary variable, R54X.

With the construction of the matrix the cluster analysis methods can be applied objectively to this set of respondents, putting the 101 respondents into a preset number of groups. The most straightforward cluster is one big cluster called the total panel. It is acceptable but not instructive to put all 101 respondents into the analysis. But what about two clusters or three clusters? At that point, the world of clustering offers up different methods. The underlying commonality of all the methods is that, in the end.

1. Each of the items (viz., each person) is assigned to only one group.
2. The set of groups comprises all of the people.
3. The people or items in a cluster, in a group, are "similar to each other" in terms of the elements on which they generate high coefficients.
4. The centroids, the average or center points of the different groups or clusters are as different as possible or at least quite different since the word "possible" is not easily definable here.

The wonderful thing about the clustering is the ease with which we can see new patterns emerging. What seemed to be that *blooming buzzing confusion* for the total panel, data without a story, emerges as data with a story.

Table 6.16 shows the results for the total panel versus the two mind-sets and versus the three mind-sets. The table is sorted by the coefficients for the three mind-sets. Each of the three blocks of coefficients for the mind-set shows the strong performing elements for that mind-set. When the same element performs strongly in both two and three mind-sets, the elements are repeated. Due to the strong performance of many elements when we made the "cut-off" 21 or higher, this time, we "raise the bar" by increasing the cutoff to 25 for the table we present. For the automated summarization of the two mind-sets by Idea Coach (Artificial Intelligence), we leave the summarization cutoff as 21.

Table 6.16 Coefficients for the total panel, for the two-mind-set solution, and for the three-mind-set solution. (MS = mind-set)

		Total	MS 1 of 2	MS 2 of 2	MS 1 of 3	MS 2 of 3	MS 3 of 3
	Base (number of respondents in this group)	101	47	54	26	35	40
	Mind-Set 1 of 2: Sustainable traffic and environmental conservation						
C3	Orient houses away from potential noise sources, such as major roads or industrial areas, to minimize noise impact on residents.	25	30	21	29	20	27
D3	Establish designated wildlife corridors throughout the golf course to maintain connectivity between habitats.	20	27	13	20	15	22
D1	Create nesting areas, bird boxes, and other structures to encourage wildlife to thrive within the golf course environment.	21	25	17	20	17	24
	Mind-Set 2 of 2 – Environmental conservation and community well being						
A3	Introduce measures to minimize vehicle collisions with wildlife, such as wildlife crossings and signage.	22	15	28	12	29	23
A1	Implement habitat restoration and enhancement programs to support local wildlife populations.	22	16	27	9	29	27
B4	Promote flexible working hours to stagger peak travel times and reduce congestion during rush hour.	20	15	25	25	33	6
	Elements which fail to meet the stringent criterion of coefficient 25+ for the two-mind-set solution						
C4	Establish a community noise complaint resolution process, where residents can voice their concerns, and ensure that their issues are addressed promptly and effectively.	23	21	24	20	23	24
C2	Provide a dedicated hotline or communication channel for residents to report noise-related issues, ensuring a swift response to address concerns.	23	23	23	22	24	24
A2	Encourage responsible waste management practices to prevent pollution of local ecosystems.	21	20	23	16	28	20
B1	Collaborate with local transportation agencies and traffic management departments to analyze traffic patterns and identify potential bottlenecks.	21	20	22	28	28	9
B1	Collaborate with local transportation agencies and traffic management departments to analyze traffic patterns and identify potential bottlenecks.	21	20	22	28	28	9
B2	Implement smart traffic signal systems that can adjust timings based on congestion levels to optimize traffic flow.	21	21	21	22	24	16
A4	Establish buffer zones around construction sites to prevent disturbances to wildlife.	18	15	21	12	26	17
C1	Implement physical barriers or soundproofing measures, such as temporary walls or fences, to minimize noise pollution during construction.	21	24	19	19	21	23
D4	Install bird-friendly designs in golf course structures, such as windows with minimal reflection or visual cues to minimize bird collisions.	21	23	19	17	20	24
B3	Develop a comprehensive transportation demand management plan that encourages carpooling, ridesharing, and the use of electric vehicles.	21	24	18	28	25	11
D2	Plant native vegetation and trees to enhance natural habitats and provide food and shelter for wildlife.	17	20	14	16	13	21

Although we used a clustering procedure that considered only the coefficients' mathematical properties, a delightfully coherent story emerges, especially for the three-mind-set solution. A less coherent story emerges when we look at two mind-sets. The reason is simple and generalizable.

The reality of clustering is that we can sort the elements better when we create more clusters or mind-sets with smaller numbers of residents in each cluster. Less thinking is involved, and realistically, there is more potential clarity when we sort these 16 ideas or messages into three mind-sets. With two mind-sets, we end up with some ideas performing well in two mind-sets. If we move to three mind-sets we may end up with clearly different groups, but we lose the power to generalize. Two and three mind-sets allow us to see "mega-patterns." Eight mind-sets allow clearer clusters, but we miss the "big picture."

Step 12: "How well did we perform in this study?" A continuing issue in research revolves around the issue of "how good is the research?" We are all familiar with the formal or informal grading of researchers, viz., the opinion that "*so and so is a good researcher*," or that "*so and so's research is mundane, mediocre*," or in the worst cases "*so and so's research is simply irrelevant, and as a researcher so and so just does not 'cut the mustard*." These are cruel but part of the "system of being human." We make these judgments again and again. The judgments can influence the career of a new researcher, and certainly ends up affecting how we think about an established researcher.

Can we measure the "goodness" of a single piece of research, such as the study we just presented here and dissected into its essentials? The topic might be interesting to some and boring to others. The reality of "what is good research" may not be truly definable to the agreement of everyone who does research. In the words of Professor S. S. Stevens of Harvard University, often quoted here and, with full disclosure, doctoral professor of author Moskowitz, "*validity is a matter of opinion*."

The issue is that we are dealing with real, "sloppy," socially meaningful topics here, not with studies that are conducted in the splendid isolation of the academic ivory tower. We must develop a way to assess the importance of the topic, or at the least the solidity of the research, a method independent of the vagaries of judgment and the ever-present annoying, possibly destructive, and certainly hindering biases of professionals in the field. Those vagaries are a source of noise, the biases a source of irrelevant bloviating, or so it seems to author Moskowitz.

An alternative approach is presented in Tables 6.14). The approach is called the Index of Divergent Thought (IDT) (Abdulla Alabbasi et al. 2021; Runco and Acar 2019). A caveat is appropriate here. The IDT is a suggested approach offered as an operationally defined way to measure the goodness of research. The IDT is not written in stone. It is simply a working method to compare studies.

The father of the divergent thinking approach is considered J.P. Guilford. In his monumental book "The Nature of Human Intelligence," (Guilford 1967) he presents an encyclopedic framework for comprehending human intelligence. Guilford goes beyond the traditional notion of intelligence as a single, general ability, thus greatly expanding the conventional view of intelligence. Guilford proposed a multifaceted model known in the literature as the Structure of Intellect (SOI) model, which identifies a broad range of mental abilities.

The SOI model organizes intelligence into three dimensions:

1. Operations (the processes used in thinking)
2. Contents (the type of information or material thought about)
3. Products (the form in which information is processed)

This categorization results in a three-dimensional configuration with many distinct facets, indicating that intelligence is composed of numerous specific abilities rather than a single general one. Guilford's work emphasizes the importance of creativity and divergent thinking as aspects of intelligence that are as important as convergent thinking and general problem-solving ability. He distinguishes between various thinking processes and encourages a broader understanding of intelligence, including abilities like spatial reasoning, memory recall, and creative generation.

Guilford's oeuvre significantly changed the approach to understanding cognitive abilities, challenged and expanded the traditional views of intelligence, and has had a lasting impact on intelligence research, education, psychological testing, and understanding cognitive abilities in miscellaneous disciplines.

With this recognition that we are looking both for strong ideas as well as for unique ideas, we can create a system based upon Guilford's stimulating ideas. We know that the data from the study comprises the set of positive coefficients for the total panel, for the two mind-sets, and then for the three mind-sets. The underlying organizing principle is that the coefficients for the elements measure the strength of performance (Caliendo and Rodríguez 2023; Baer 2014). Ideally, we would like to see each mind-set generate some strong-performing elements and other weak-performing elements, as well as see that the strong-performing elements generate very high coefficients (e.g., around 25–30). We end up with problem solving through high coefficients of elements. We end up with divergent thinking through the mind-sets.

We know the following:

1. When all the coefficients are low, we don't have any elements which solved problems.
2. When there are no mind-sets or when the mind-sets look alike in terms of strong performing coefficients, we may have solved the problem, but we don't have divergent thinking.
3. Somewhere in there is the idea point: we have mindsets with high coefficients in one mind-set and low coefficients in another. This pattern suggests that we have both problem solving and divergent thinking.

Let us now simulate what we might get if we were to have strong differentiation among mind-sets.

Table 6.17 shows the computation for simulated results assuming that there is VERY STRONG segmentation in the study so that some elements are given coefficients of 33, and other elements are given coefficients of 0. Furthermore, the coefficients are assigned in such a way that there is obvious and clear segmentation.

Table 6.17 Data pattern and calculation of the IDT (Index of Divergent Thought) for the current elements, but with simulated data, which show maximum possible level of segmentation

	Total	Mind-set 1 of 2	Mind-set 2 of 2	Mind-set 1 of 3	Mind-set 2 of 3	Mind-set 3 of 3
Base (number of respondents in this group)	**101**	**50**	**51**	**33**	**34**	**34**
Implement habitat restoration and enhancement programs to support local wildlife populations.	10	33	0	33	0	0
Encourage responsible waste management practices to prevent pollution of local ecosystems.	15	33	0	33	0	0
Introduce measures to minimize vehicle collisions with wildlife, such as wildlife crossings and signage.	20	33	0	33	0	0
Establish buffer zones around construction sites to prevent disturbances to wildlife.	10	33	0	33	0	0
Collaborate with local transportation agencies and traffic management departments to analyze traffic patterns and identify potential bottlenecks.	15	33	0	0	33	0
Implement smart traffic signal systems that can adjust timings based on congestion levels to optimize traffic flow.	20	33	0	0	33	0

This would be an ideal result, viz., strong performing elements and clear segmentation.

The key number in Table 6.17 is the Index of Divergent Thought, IDT. For Table 6.17, the IDT is around 78. This number, 78, becomes an anchor point. It is an approximate reason for a really strong segmentation pattern. The number 78 does not help us interpret the mind-sets but simply gives us a sense of what a strong performing pattern would look like. Parenthetically, if the pattern were uninterpretable, we would discard the clustering as being non-instructive.

Table 6.18 shows the same set of elements, this time with the actual coefficients from the study results. The IDT is higher, 86 rather than 78. We can say that the value 86 comes from the fact that the segmentation does not strongly separate the groups. The patterns are still interpretable as we will see from the AI summarization, but the segmentation could be stronger. The value 78 or so seems to be the area of optimal clustering, especially when the clustering or segmentation shows different types of elements appealing to the mind-sets, as it did in Table 6.17.

Step 13: Summarizing the mind-sets using Chat GPT3.5. One of the benefits of Mind Genomics is the ability to invoke AI to help craft a story about the strong-performing elements. Table 6.19 summarizes key information about the two mind-sets provided by AI (Idea Coach). The queries are those given to Idea Coach, which returns suggestions based only on the elements with coefficients of +21 or higher.

Table 6.18 Data pattern and calculation for the current elements, and the actual coefficients

	Total	Mind-set 1 of 2	Mind-set 2 of 2	Mind-set 1 of 3	Mind-set 2 of 3	Mind-set 3 of 3
Base (number of respondents in this group)	**101**	**47**	**54**	**26**	**35**	**40**
Implement habitat restoration and enhancement programs to support local wildlife populations.	22	16	27	9	29	27
Encourage responsible waste management practices to prevent pollution of local ecosystems.	21	20	23	16	28	20
Introduce measures to minimize vehicle collisions with wildlife, such as wildlife crossings and signage.	22	15	28	12	29	23
Establish buffer zones around construction sites to prevent disturbances to wildlife.	18	15	21	12	26	17
Collaborate with local transportation agencies and traffic management departments to analyze traffic patterns and identify potential bottlenecks.	21	20	22	28	28	9
Implement smart traffic signal systems that can adjust timings based on congestion levels to optimize traffic flow.	21	21	21	22	24	16
Develop a comprehensive transportation demand management plan that encourages carpooling, ridesharing, and the use of electric vehicles.	21	24	18	28	25	11
Promote flexible working hours to stagger peak travel times and reduce congestion during rush hour.	20	15	25	25	33	6
Implement physical barriers or soundproofing measures, such as temporary walls or fences, to minimize noise pollution during construction.	21	24	19	19	21	23

When no elements are in a subgroup of 21 or higher, the AI moves to the next subgroup.

We reproduce the output here to give an idea of the breadth of information available from the Mind Genomics platform. In future versions, enhancing the AI "analyses" or "summarization" of the results with other queries should be possible. However, we might wish to avoid the potentially overwhelming "richness" of interpretation open to us due to having cognitively meaningful elements.

Step 14: Assigning new people to mind-sets. How can we generalize these results? In other words, we have relatively detailed information about how people think regarding this micro-topic of builder concession. The study might well have been conducted in a single geographic region, among a limited number of individuals with high income. The output would still reveal several mind-sets. In the words of the late singer of the 1950s, Peggy Lee, "*Is that all there is*?"

Table 6.19 Idea Coach (AI based) summarization the strong elements from the two-mind-set solution

DEPENDENT VARIABLE = 54X
GOOD BUILD CONCESSIONS
Mind-set 1 of 2—Top (R54X > = 21)
Sustainable Transportation and Environmental Conservation
Interested in:
Implement smart traffic signal systems that can adjust timings based on congestion levels to optimize traffic flow; Develop a comprehensive transportation demand management plan that encourages carpooling, ride-sharing, and the use of electric vehicles; Implement physical barriers or soundproofing measures, such as temporary walls or fences, to minimize noise pollution during construction; Provide a dedicated hotline or communication channel for residents to report noise-related issues, ensuring a swift response to address concerns; Orient houses away from potential noise sources, such as major roads or industrial areas, to minimize noise impact on residents.; Establish a community noise complaint resolution process, where residents can voice their concerns, and ensure that their issues are addressed promptly and effectively; Create nesting areas, bird boxes, and other structures to encourage wildlife to thrive within the golf course environment; Establish designated wildlife corridors throughout the golf course to maintain connectivity between habitats; Install bird-friendly designs in golf course structures, such as windows with minimal reflection or visual cues to minimize bird collisions;
Create a label for this segment:
Sustainable Transportation and Environmental Conservation
Describe this segment:
This segment is focused on improving transportation systems and reducing noise pollution in residential areas. They propose implementing smart traffic signal systems to optimize traffic flow, developing transportation demand management plans to encourage carpooling and the use of electric vehicles, and implementing measures to minimize noise pollution during construction. They also emphasize the importance of providing a dedicated hotline for residents to report noise-related issues and ensuring a prompt response to address concerns. Furthermore, they suggest orienting houses away from potential noise sources, establishing a community noise complaint resolution process, and creating nesting areas and wildlife corridors within golf courses to promote the thriving of wildlife. They also emphasize the installation of bird-friendly designs in golf course structures to prevent bird collisions.
Describe the attractiveness of this segment as a target audience:
Based on the given attributes, this segment appears to be environmentally conscious and concerned about minimizing the negative impacts of various activities on the environment and people's quality of life. They are interested in implementing smart traffic signal systems to optimize traffic flow and reduce congestion. They also prioritize developing a comprehensive transportation demand management plan that encourages sustainable practices like carpooling, ride-sharing, and the use of electric vehicles.
In addition, this segment shows a strong commitment to minimizing noise pollution during construction by implementing physical barriers or soundproofing measures. They value open communication and swift response to address noise-related issues, as demonstrated by the dedicated hotline or communication channel for residents to report concerns and the establishment of a noise complaint resolution process.

(continued)

Table 6.19 (continued)

Moreover, this segment also highlights their concern for wildlife and their desire to create a harmonious environment for them within the golf course. They emphasize the importance of establishing wildlife corridors to maintain connectivity between habitats and implementing bird-friendly designs to minimize bird collisions.
Overall, this segment can be considered highly attractive as a target audience due to their strong environmental consciousness, commitment to sustainable practices, and consideration for both human and wildlife well-being. They are likely to be receptive to initiatives and policies aimed at improving the environment and minimizing negative impacts.
Explain why this segment might not be attractive as a target audience:
This segment might not be attractive as a target audience because it consists of individuals who are primarily concerned with traffic management, noise pollution, and wildlife conservation in the context of a golf course environment. These issues and solutions may not be applicable or relevant to a wide range of individuals or businesses outside of the specific setting of a golf course. Thus, targeting this segment may not result in a significant reach or impact on a broader audience. Additionally, implementing some of these solutions may require significant resources, making it less appealing for certain organizations or individuals to engage with this segment as a target audience.
List what is missing or should be known about this segment, in question form:
1. How effective have smart traffic signal systems been in optimizing traffic flow in other areas?
2. Have transportation demand management plans successfully incentivized carpooling, ride-sharing, and electric vehicle usage in similar contexts?
3. What specific noise levels are considered acceptable during construction and how do they compare to existing regulations?
4. How accessible and user-friendly is the dedicated hotline or communication channel for residents to report noise-related issues?
5. What measures or considerations have been taken to ensure that houses are effectively oriented away from potential noise sources?
6. How successful has the community noise complaint resolution process been in addressing residents' concerns?
7. Have nesting areas, bird boxes, and other wildlife structures been previously implemented in other golf courses, and what has been their impact on wildlife populations?
8. Are designated wildlife corridors currently present in the golf course and if so, how have they contributed to maintaining habitat connectivity?
9. What bird-friendly designs have been proven effective in minimizing bird collisions in golf course structures?
10. What other potential environmental impacts, if any, could arise from implementing these measures and how can they be mitigated?
List and briefly describe attractive new or innovative products, services, experiences, or policies for this segment:
Some attractive new or innovative products, services, experiences, or policies for this segment could include:
1. Smart Traffic Signal Systems: Implementing smart traffic signal systems that can adjust timings based on congestion levels can help optimize traffic flow and reduce traffic congestion. This would lead to a smoother and more efficient transportation system.
2. Transportation Demand Management Plan: Developing a comprehensive transportation demand management plan that encourages carpooling, ride-sharing, and the use of electric vehicles can help reduce traffic congestion and promote sustainable transportation options.

(continued)

Table 6.19 (continued)

3. Noise Pollution Mitigation Measures: Implementing physical barriers or soundproofing measures, such as temporary walls or fences, during construction can help minimize noise pollution and reduce its impact on residents.
4. Dedicated Hotline for Noise-related Issues: Providing a dedicated hotline or communication channel for residents to report noise-related issues ensures a swift response to address concerns. This would help improve resident satisfaction and address noise-related problems effectively.
5. Strategic House Orientation: Orienting houses away from potential noise sources, such as major roads or industrial areas, can minimize noise impact on residents and create a quieter living environment.
6. Community Noise Complaint Resolution Process: Establishing a community noise complaint resolution process where residents can voice their concerns and ensure that their issues are addressed promptly and effectively can help improve resident satisfaction and maintain a harmonious community.
7. Wildlife Conservation Efforts: Creating nesting areas, bird boxes, and other structures to encourage wildlife to thrive within the golf course environment can enhance biodiversity and create a more attractive and eco-friendly golfing experience.
8. Wildlife Corridors: Establishing designated wildlife corridors throughout the golf course to maintain connectivity between habitats ensures the preservation of wildlife and their natural movement patterns within the golf course area.
9. Bird-friendly Golf Course Structures: Installing bird-friendly designs in golf course structures, such as windows with minimal reflection or visual cues to minimize bird collisions, demonstrates a commitment to wildlife conservation and supports the well-being of local bird populations.
Which messages will interest this segment?
1. “Implement smart traffic signal systems that can adjust timings based on congestion levels to optimize traffic flow.”
2. “Develop a comprehensive transportation demand management plan that encourages carpooling, ride-sharing, and the use of electric vehicles.”
3. “Implement physical barriers or soundproofing measures, such as temporary walls or fences, to minimize noise pollution during construction.”
4. “Provide a dedicated hotline or communication channel for residents to report noise-related issues, ensuring a swift response to address concerns.”
5. “Orient houses away from potential noise sources, such as major roads or industrial areas, to minimize noise impact on residents.”
6. “Establish a community noise complaint resolution process, where residents can voice their concerns, and ensure that their issues are addressed promptly and effectively.”
7. “Create nesting areas, bird boxes, and other structures to encourage wildlife to thrive within the golf course environment.”
8. “Establish designated wildlife corridors throughout the golf course to maintain connectivity between habitats.”
9. “Install bird-friendly designs in golf course structures, such as windows with minimal reflection or visual cues to minimize bird collisions.”
GOOD BUILDER CONCESSIONS
Mind-set 2 of 2—Top (R54X > = 21)
Environmental Conservation and Community Well-being
Interested in:

(continued)

Table 6.19 (continued)

Implement habitat restoration and enhancement programs to support local wildlife populations; Encourage responsible waste management practices to prevent pollution of local ecosystems; Introduce measures to minimize vehicle collisions with wildlife, such as wildlife crossings and signage; Establish buffer zones around construction sites to prevent disturbances to wildlife; Collaborate with local transportation agencies and traffic management departments to analyze traffic patterns and identify potential bottlenecks; Implement smart traffic signal systems that can adjust timings based on congestion levels to optimize traffic flow; Promote flexible working hours to stagger peak travel times and reduce congestion during rush hour; Provide a dedicated hotline or communication channel for residents to report noise-related issues, ensuring a swift response to address concerns; Orient houses away from potential noise sources, such as major roads or industrial areas, to minimize noise impact on residents; Establish a community noise complaint resolution process, where residents can voice their concerns, and ensure that their issues are addressed promptly and effectively.
Create a label for this segment:
Environmental Conservation and Community Well-being
Describe this segment:
This segment is focused on implementing measures to support and enhance local wildlife populations, prevent pollution of local ecosystems, and minimize disruptions to wildlife caused by human activities. It also emphasizes collaboration with transportation agencies and traffic management departments to analyze and optimize traffic flow, reduce congestion during rush hour, and minimize the impact of vehicle collisions with wildlife. Additionally, this segment aims to address noise-related issues and concerns by providing a dedicated communication channel for residents to report noise problems and establishing a community complaint resolution process. Overall, this segment is dedicated to creating a more sustainable and environmentally friendly community that prioritizes the well-being of both wildlife populations and residents.
Describe the attractiveness of this segment as a target audience:
This segment of individuals and communities demonstrates a strong commitment to environmental conservation and the well-being of local wildlife populations. They prioritize sustainable practices and demonstrate a willingness to take action in order to preserve and enhance the natural habitats and ecosystems in their area. Their interest in implementing habitat restoration and enhancement programs, responsible waste management practices, and measures to minimize vehicle collisions with wildlife indicates a deep concern for the protection and conservation of local wildlife populations.
Furthermore, their collaboration with transportation agencies and traffic management departments shows their proactive approach in identifying and addressing traffic congestion issues, reflecting their commitment to reduce pollution and improve traffic flow. Their promotion of flexible working hours to reduce congestion during rush hour exemplifies their consideration for the overall welfare and quality of life for residents in their community.
The dedication to addressing noise-related issues through prompt response and resolution processes demonstrates their commitment to ensuring a peaceful and harmonious living environment for residents. Additionally, their efforts to orient houses away from potential noise sources, such as major roads or industrial areas, further indicates their focus on providing a high quality of life for residents and minimizing the negative impacts of noise pollution.
Overall, this target audience is highly attractive due to their strong sense of environmental stewardship, proactive problem-solving approach, and commitment to finding sustainable solutions for the benefit of both wildlife populations and residents.
Explain why this segment might not be attractive as a target audience:

(continued)

Table 6.19 (continued)

This segment might not be attractive as a target audience because their primary interests and concerns are focused on wildlife conservation, habitat restoration, waste management, and transportation and traffic-related issues. These interests may not align with the objectives or products/services that a company or organization is trying to promote. Additionally, the segment may have specific needs and preferences that may not be easily met or fulfilled by the company, therefore making it less of a viable target audience.
List what is missing or should be known about this segment, in question form:
1. What are the specific wildlife populations that are in need of habitat restoration and enhancement programs?
2. Are there specific pollution sources or waste management practices that are most prevalent and need to be targeted?
3. Which specific areas or highways have a high incidence of wildlife-vehicle collisions and require wildlife crossings and signage?
4. What level of disturbance does the wildlife experience from construction sites, and what types of buffer zones would be most effective?
5. Have traffic patterns and bottlenecks already been analyzed, or is further analysis needed to identify potential solutions?
6. Are there specific peak travel times that need to be addressed through the implementation of flexible working hours?
7. What are the main noise-related issues commonly experienced by residents in the area?
8. Are there specific noise sources, besides major roads, and industrial areas, that also need to be considered when orienting houses?
9. Is there an existing process for resolving noise complaints or is the establishment of a community noise complaint resolution process new?
10. What are the potential barriers or challenges in implementing these measures, such as financial constraints or community resistance?
List and briefly describe attractive new or innovative products, services, experiences, or policies for this segment:
1. Wildlife-friendly landscaping services: Offer landscaping services that focus on creating and maintaining habitats suitable for local wildlife. This can include planting native vegetation, providing nesting areas, and installing bird feeders or bat boxes.
2. Waste reduction and recycling incentive programs: Implement programs that incentivize responsible waste management practices, such as offering discounts or rewards for recycling or using reusable products. This can help reduce pollution in local ecosystems by minimizing waste and promoting sustainable practices.
3. Wildlife-friendly infrastructure design: Develop innovative infrastructure designs that minimize vehicle collisions with wildlife. This can include the installation of wildlife crossings, signage, and fencing to guide animals away from busy roads and highways.
4. Noise-cancelling technologies for residential areas: Introduce innovative noise-cancelling technologies for homes located near major roads or industrial areas. This can help minimize the impact of noise pollution on residents and improve their quality of life.
5. Community-based traffic management platforms: Establish digital platforms or mobile apps that allow residents to monitor and report traffic-related issues in real-time. This can provide valuable data for transportation agencies and help identify potential bottlenecks or areas that require traffic flow optimization.

(continued)

Table 6.19 (continued)

6. Wildlife-friendly construction practices: Encourage developers and construction companies to adopt wildlife-friendly practices by establishing buffer zones around construction sites. This can help minimize disturbances to local wildlife habitats during the construction process.
7. Eco-friendly transportation solutions: Introduce innovative transportation options such as electric or hybrid vehicles, bike-sharing programs, or efficient public transit systems. This can help reduce congestion, pollution, and noise levels in the community.
8. Smart home technology for energy and resource conservation: Promote the use of smart home devices and technologies that allow residents to monitor and optimize their energy consumption. This can help individuals reduce their ecological footprint and contribute to a more sustainable community.
9. Nature education and ecotourism experiences: Develop nature education programs and ecotourism experiences that promote environmental awareness and appreciation for local wildlife. This can include guided tours, workshops, or volunteer opportunities to engage residents and visitors with the natural beauty of the area.
10. Sustainable urban planning policies: Advocate for sustainable urban planning policies that prioritize green spaces, biodiversity conservation, and sustainable development. This can include implementing green building codes, preserving natural habitats, and incorporating renewable energy sources in urban infrastructure.
Which messages will interest this segment?
1. "Implement habitat restoration and enhancement programs to support local wildlife populations."
2. "Encourage responsible waste management practices to prevent pollution of local ecosystems."
3. "Introduce measures to minimize vehicle collisions with wildlife, such as wildlife crossings and signage."
4. "Establish buffer zones around construction sites to prevent disturbances to wildlife."
5. "Collaborate with local transportation agencies and traffic management departments to analyze traffic patterns and identify potential bottlenecks."
6. "Implement smart traffic signal systems that can adjust timings based on congestion levels to optimize traffic flow."
7. "Promote flexible working hours to stagger peak travel times and reduce congestion during rush hour."
8. "Provide a dedicated hotline or communication channel for residents to report noise-related issues, ensuring a swift response to address concerns."
9. "Orient houses away from potential noise sources, such as major roads or industrial areas, to minimize noise impact on residents."
10. "Establish a community noise complaint resolution process, where residents can voice their concerns, and ensure that their issues are addressed promptly and effectively."
These messages will interest the segment interested in environmental conservation, wildlife protection, traffic management, and noise pollution mitigation.

Are the results from these Mind Genomics studies destined to simply constitute another momentary "picture" of the mind of people concerning a specific topic, in a place and time, fixed by the "accident" of when the work was done, why the work was done, and with whom was the work done? At some later date, the historian of Mind Genomics might well assemble these disparate studies about how people think and attempt to find a pattern in the data, connecting these wildly different types of studies. This path describes a great deal of meta-data analysis, looking for general patterns from disparate studies about the same general topic.

An alternative opportunity consists of creating a short, easy-to-administer "tool," which can quickly assign a new person to one of the two mind-sets. When the three-mind-set solution is selected, the same tool can assign a person to one of the three mind-sets. The mathematics are the same. Quite simply, the PVI system tries out all combinations of elements until it finds that combination of elements, which most closely reproduces the proper assignment of an individual. The approach works in a Monte-Carlo fashion (Ferkingstad et al. 2015), slightly perturbing the patterns to add "noise." The objective is to create a system that can perform reasonably well in the presence of noisy data.

Like the Mind Genomics platform, BimiLeap, the PVI (personal viewpoint identifier), has been "templated" to make the creation of the PVI easy for any specific data set emerging from a study. The templates appear in Figs. 6.11a, 6.11b, and 6.11c, respectively .

Figure 6.11a shows a screenshot of the "PVI Booklet Management" page. The new study is called Neighborhood 1 and appears as the first row of studies. When the researcher begins the process, the researcher can either start a new booklet, like is done here or add to the booklet. Adding to the booklet occurs in cases where the researcher wants to use several PVI's with a single respondent to create a "mind-print" for a topic. An example is medicine, where the PVI is created for different situations in a doctor or hospital visit, such as how to be told distressing news, how to be motivated to comply with the doctor's suggestions, and perhaps one or two

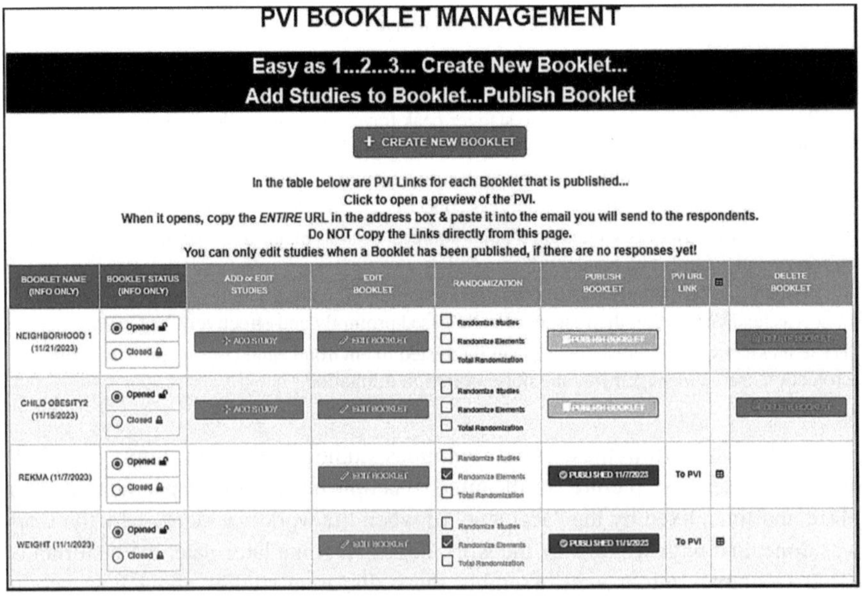

Fig. 6.11a PVI booklet manager. The new study is called Neighborhood 1

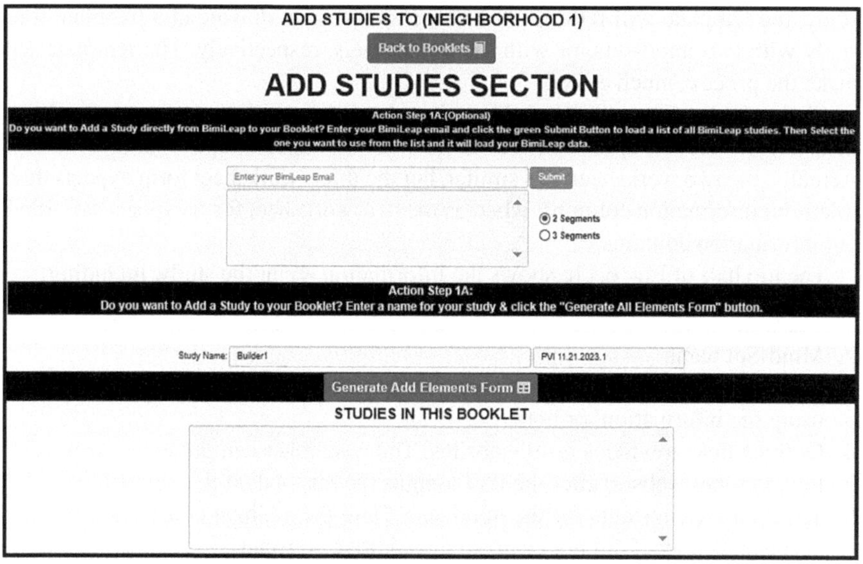

Fig. 6.11b Request to add a new study

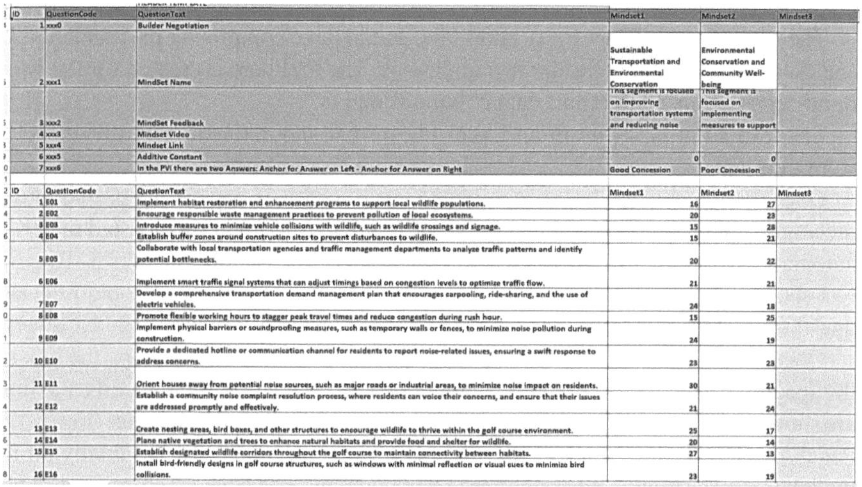

ID	QuestionCode	QuestionText	Mindset1	Mindset2	Mindset3
1	xxx0	Builder Negotiation			
2	xxx1	MindSet Name	Sustainable Transportation and Environmental Conservation	Environmental Conservation and Community Well-being	
3	xxx2	MindSet Feedback	This segment is focused on improving transportation systems and reducing noise	This segment is focused on implementing measures to support	
4	xxx3	Mindset Video			
5	xxx4	Mindset Link			
6	xxx5	Additive Constant	0	0	
7	xxx6	In the PVI there are two Answers: Anchor for Answer on Left - Anchor for Answer on Right	Good Concession	Poor Concession	

ID	QuestionCode	QuestionText	Mindset1	Mindset2	Mindset3
1	E01	Implement habitat restoration and enhancement programs to support local wildlife populations.	16	27	
2	E02	Encourage responsible waste management practices to prevent pollution of local ecosystems.	20	23	
3	E03	Introduce measures to minimize vehicle collisions with wildlife, such as wildlife crossings and signage.	15	28	
4	E04	Establish buffer zones around construction sites to prevent disturbances to wildlife.	15	21	
5	E05	Collaborate with local transportation agencies and traffic management departments to analyze traffic patterns and identify potential bottlenecks.	20	22	
6	E06	Implement smart traffic signal systems that can adjust timings based on congestion levels to optimize traffic flow.	21	21	
7	E07	Develop a comprehensive transportation demand management plan that encourages carpooling, ride-sharing, and the use of electric vehicles.	24	18	
8	E08	Promote flexible working hours to stagger peak travel times and reduce congestion during rush hour.	15	25	
9	E09	Implement physical barriers or soundproofing measures, such as temporary walls or fences, to minimize noise pollution during construction.	24	19	
10	E10	Provide a dedicated hotline or communication channel for residents to report noise-related issues, ensuring a swift response to address concerns.	23	23	
11	E11	Orient houses away from potential noise sources, such as major roads or industrial areas, to minimize noise impact on residents.	30	21	
12	E12	Establish a community noise complaint resolution process, where residents can voice their concerns, and ensure that their issues are addressed promptly and effectively.	21	24	
13	E13	Create nesting areas, bird boxes, and other structures to encourage wildlife to thrive within the golf course environment.	25	17	
14	E14	Plane native vegetation and trees to enhance natural habitats and provide food and shelter for wildlife.	20	14	
15	E15	Establish designated wildlife corridors throughout the golf course to maintain connectivity between habitats.	27	13	
16	E16	Install bird-friendly designs in golf course structures, such as windows with minimal reflection or visual cues to minimize bird collisions.	23	19	

Fig. 6.11c The worksheet, filled out, read to be "copied into the PVI"

other things. All four would be presented in a randomized sequence. The order, etc., would be maintained by having the PVI's for all four studies located in a single book.

Figure 6.11a; The PVI booklet manager. The new study is called Neighborhood 1. The study is the first study on the screen shot.

Figure 6.11b shows the request to add studies to the newly created booklet. The templated figure leads the user through the steps. Once the user has named the study, the template will return with instructions how to download a template for a study with two mind-sets, or with three mind-sets, respectively. The template will make the process much easier.

Figure 6.11c shows the worksheet, downloaded from the website as two Excel files, a worksheet set up for two mind-sets or a worksheet set up for three mind-sets. Actually, the two worksheets are similar, but the three-worksheet form expects three filled-out information columns, whereas the two-worksheet form expects two filled-out information columns.

The top half of Fig. 6.11c shows the information about the study, including:

1. Name of study
2. Mind-Set name
3. Mind-Set feedback, either to the respondent taking the PVI, to the professional using the information, or both
4. Options links to videos or to websites. The researcher can guide the respondent to the proper website after the PVI assigns the respondent to a mind-type
5. The additive constants for the mind-sets. Here the additive constants are both 0 because the model did not estimate the additive constant
6. Anchors for the two-point questions. The respondent will be presented with six elements in the form of questions, and with two answers, a top anchor and a bottom anchor respectively

The bottom half of Fig. 6.11c shows the actual data, comprising the element text and then the coefficients for the two mind-sets. The PVI here is created for the two-mind-set solution, so the third data column is empty.

The PVI form is set up in an Excel file format and automatically downloaded from the site.

Once the Excel worksheet is completed, each section is copied and pasted into a rectangle of the same color. When the setup is deemed "correct," the pasted material automatically populates the sheet. The typical time to create the sheet is 5–10 min. Not shown are the 16 additional questions, the so-called specialty, or topic-relevant questions that the researcher can create, each requiring a selection from 2–4 answers.

Once the PVI has been created and registered, the researcher can copy the link to the specific PVI and send that link to many people. The people who receive the link are shown the PVI itself, beginning with information about the respondent (Fig. 6.12, Panel A), then six question PVI for this study (Fig. 6.12, Panel B), and finally information about the mind-set to which the respondent has just been assigned (Fig. 6.12, Pane C) If there were two studies, the questions for both would be shown, one set after another. The PVI setup allows the researcher to randomize the order of the multiple PVI's and the order of the six questions). Furthermore, ALL the information is stored in a project database.

Figure 6.12 shows the PVI experience. Panel A shows the background information collected about the respondent. Panel B shows the actual six questions asked of the respondent, and the two standard answers. Panel C shows the feedback provider to the user. All the information is stored in the database.

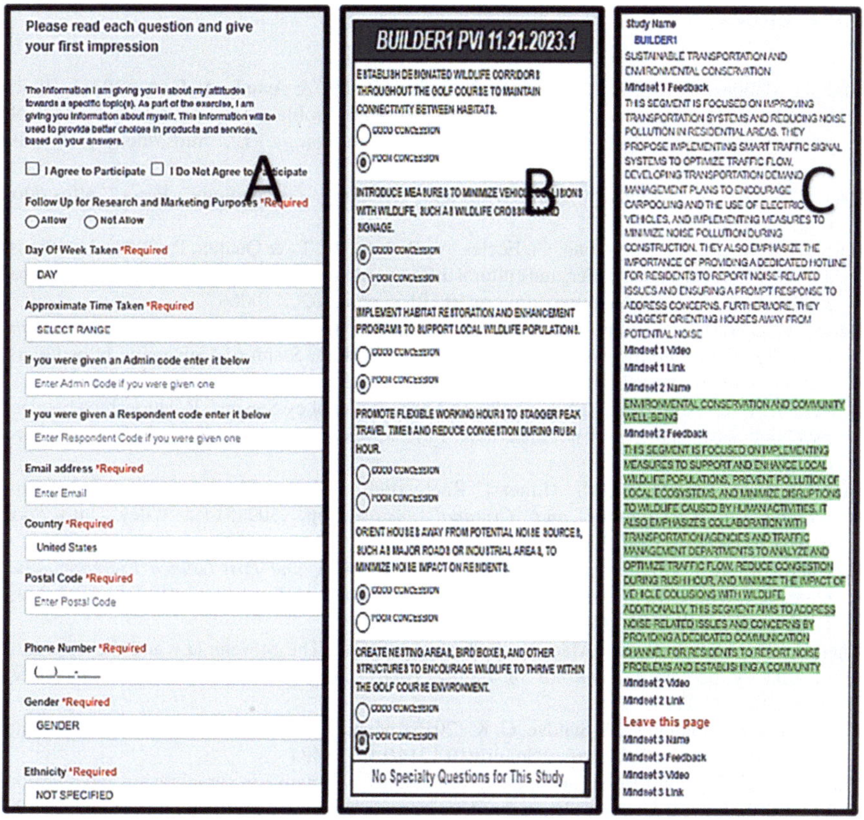

Fig. 6.12 The PVI Experience. Panel A shows the information about the respondent. Panel B shows the six questions used to assign a person to one of the two mind-sets. Panel C shows the feedback given either to the respondent or the person administering the PVI to the respondent

The PVI feature increases the sample size by including respondents who did not participate in the experiment, thus creating an open system. This powerful feature has been used successfully in several studies (Zemel et al. 2019; Zemel et al. 2018). As with any new approach, the PVI is slowly gaining traction, and creating interest. The potential for the PVI is great, as a way to understand populations of people on a variety of issues investigated through Mind Genomics. One can imagine creating ten studies of the type described in this chapter, and for each study creating a separate PVI of six questions for each study. In a matter of 3–4 min, it become possible for a person to answer the 60 questions with a simple yes/no. The happy result is the ability to understand the mind of the individual on those ten topics. The potential increases when we realize that this approach can be exported worldwide to create a database of the mind on political and social issues.

References

Abdulla Alabbasi, A. M., Reiter-Palmon, R., Sultan, Z. M., & Ayoub, A. E. A. (2021). Which Divergent Thinking Index Is More Associated With Problem Finding Ability? The Role of Flexibility and Task Nature. *Frontiers in Psychology, 12*. https://doi.org/10.3389/fpsyg.2021.671146

Baer, J. (2014). *Creativity and Divergent Thinking*. Psychology Press. https://doi.org/10.4324/9781315806785

Barreto, M., Victor, C., Hammond, C., Eccles, A., Richins, M. T., & Qualter, P. (2021). Loneliness around the world: Age, gender, and cultural differences in loneliness. *Personality and Individual Differences, 169*, 110066. https://doi.org/10.1016/j.paid.2020.110066

Beiser, F. C. (2020). Gustav Theodor Fechner. In *The Stanford Encyclopedia of Philosophy Encyclopedia of Philosophy*. Metaphysics Research Lab, Stanford University. https://plato.stanford.edu/archives/spr2020/entries/fechner/

Billock, V. A., & Tsou, B. H. (2011). To honor Fechner and obey Stevens: Relationships between psychophysical and neural nonlinearities. *Psychological Bulletin, 137*(1), 1–18. https://doi.org/10.1037/a0021394

Burton, A. L. (2021). OLS (Linear) Regression. In *The Encyclopedia of Research Methods in Criminology and Criminal Justice* (pp. 509–514). Wiley. https://doi.org/10.1002/9781119111931.ch104

Caliendo, M., & Rodríguez, D. (2023). *Divergent Thinking and Post-Launch Entrepreneurial Outcomes: Non-Linearities and the Moderating Role of Experience* (ISSN: 2365-9793; DISCUSSION PAPER SERIES).

Chen, J., Tian, S., Wang, X., Mao, Y., & Zhao, L. (2021). The Stevens law and the derivation of sensory perception. *Journal of Future Foods, 1*(1), 82–87. https://doi.org/10.1016/j.jfutfo.2021.09.004

Ferkingstad, E., Holden, L., & Sandve, G. K. (2015). Monte Carlo Null Models for Genomic Data. *Statistical Science, 30*(1). https://doi.org/10.1214/14-STS484

Fitria, T. N. (2021). QuillBot as an online tool: Students' alternative in paraphrasing and rewriting of English writing. *Englisia: Journal of Language, Education, and Humanities, 9*(1), 183–196. https://doi.org/10.22373/ej.v9i1.10233

Flatt, C., & Jacobs, R. L. (2019). Principle Assumptions of Regression Analysis: Testing, Techniques, and Statistical Reporting of Imperfect Data Sets. *Advances in Developing Human Resources, 21*(4), 484–502. https://doi.org/10.1177/1523422319869915

Galanter, E., & Pliner, P. (1974). Cross-Modality Matching of Money Against Other Continua. In *Sensation and Measurement* (pp. 65–76). Springer Netherlands. https://doi.org/10.1007/978-94-010-2245-3_6

Gofman, A., & Moskowitz, H. (2010a). Application of Isomorphic Permuted Experimental Designs in Conjoint Analysis". *Journal of Sensory Studies, 25*(1), 127–145. https://doi.org/10.1111/j.1745-459X.2009.00258.x

Gofman, A., & Moskowitz, H. (2010b). ISOMORPHIC PERMUTED EXPERIMENTAL DESIGNS AND THEIR APPLICATION IN CONJOINT ANALYSIS. *Journal of Sensory Studies, 25*(1). https://doi.org/10.1111/j.1745-459X.2009.00258.x

Gofman, A., & Moskowitz, H. R. (2010c). Isomorphic Permuted Experimental Designs And Their Application In Conjoint Analysis. *Journal of Sensory Studies, 25*(1), 127–145. https://doi.org/10.1111/j.1745-459X.2009.00258.x

Goodman, R. (2022). William James. In *The Stanford Encyclopedia of Philosophy* (Spring 202). Metaphysics Research Lab, Stanford University. https://plato.stanford.edu/archives/spr2022/entries/james/

Guilford, J. P. (1967). *The nature of human intelligence*.

Heidelberger, M. (1993). Fechner's impact for measurement theory. *Behavioral and Brain Sciences, 16*(1), 146–148. https://doi.org/10.1017/S0140525X00029368

Heidelberger, M. (1994). The Unity of Nature and Mind: Gustav Theodor Fechner's Non-Reductive Materialism. In *Romanticism in Science*.

Kodinariya, T. M., & Prashant, R. M. (2013). Review on determining number of Cluster in K-Means Clustering. *International Journal of Advance Research in Computer Science and Management Studies*, *16*. www.ijarcsms.com

Lawson, A. E. (2000). The Generality of Hypothetico-Deductive Reasoning: Making Scientific Thinking Explicit. *American Biology Teacher*, *62*(7), 482–495.

Moskowitz, H. R., Gofman, A., Beckley, J., & Ashman, H. (2006). FOUNDING A NEW SCIENCE: MIND GENOMICS. *Journal of Sensory Studies*, *21*(3), 266–307. https://doi.org/10.1111/j.1745-459X.2004.00066.x

Moskowitz, H., Wren, J., & Papajorgji, P. (2020). *Mind Genomics and the Law* (1st Editio). LAP LAMBERT Academic Publishing.

Mucherino, A., Papajorgji, P., & Pardalos, P. (2009). Clustering by k-means. In *Data Mining in Agriculture* (pp. 47–82).

OpenAI. (2023). *ChatGPT* (GPT-4) [Large language model].

Ortlieb, S. A., Kügel, W. A., & Carbon, C.-C. (2020). Fechner (1866): The Aesthetic Association Principle—A Commented Translation. *I-Perception*, *11*(3), 204166952092030. https://doi.org/10.1177/2041669520920309

Porretta, S., Gere, A., Radványi, D., & Moskowitz, H. (2019). Mind Genomics (Conjoint Analysis): The new concept research in the analysis of consumer behaviour and choice. *Trends in Food Science and Technology*, *84*(January), 29–33. https://doi.org/10.1016/j.tifs.2018.01.004

Presson, A. P., Zhang, C., Abtahi, A. M., Kean, J., Hung, M., & Tyser, A. R. (2017). Psychometric properties of the Press Ganey® Outpatient Medical Practice Survey. *Health and Quality of Life Outcomes*, 15(1), 32. https://doi.org/10.1186/s12955-017-0610-3

Rose, A. M., & Prell, A. E. (1955). Does the Punishment Fit the Crime? A Study in Social Valuation. *American Journal of Sociology*, *61*(3), 247–259. https://www.jstor.org/stable/2772137

Runco, M. A., & Acar, S. (2019). Divergent Thinking. In *The Cambridge Handbook of Creativity* (pp. 224–254). Cambridge University Press. https://doi.org/10.1017/9781316979839.013

Ryan, T. P., & Morgan, J. P. (2007). Modern Experimental Design. *Journal of Statistical Theory and Practice*, 1(3–4), 501–506. https://doi.org/10.1080/15598608.2007.10411855

Smith, P. B., & Bond, M. H. (2019). Cultures and Persons: Characterizing National and Other Types of Cultural Difference Can Also Aid Our Understanding and Prediction of Individual Variability. *Frontiers in Psychology*, *10*. https://doi.org/10.3389/fpsyg.2019.02689

Stevens, S. S. (1975). *Psychophysics: Introduction to its perceptual, neural, and social prospects*. John Wiley & Sons Inc.

Zdaniuk, B. (2014). Ordinary Least-Squares (OLS) Model. In *Encyclopedia of Quality of Life and Well-Being Research* (2014. Edit, pp. 4515–4517). Springer Netherlands. https://doi.org/10.1007/978-94-007-0753-5_2008

Zemel, R., Gan Choudhuri, S., Gere, A., Upreti, H., Deitel, Y., Papajorgji, P., & Moskowitz, H. (2019). Consumers, and Dairy: Applying Artificial Intelligence, Mind Genomics, and Predictive Viewpoint Typing. In *Handbook of Modern Dairy Science and Technology*. IntechOpen. https://doi.org/10.5772/intecopen.85532

Zemel, R., Gere, A., Papajorgji, P., Zemel, G., & Moskowitz, H. (2018). Uncovering Consumer Mind-sets Regarding Raw Beverages. *Food and Nutrition Sciences*, 9(3), 259–267. https://doi.org/10.4236/fns.2018.93020

Zimmerman, C., & Croker, S. (2014). A Prospective Cognition Analysis of Scientific Thinking and the Implications for Teaching and Learning Science. *Journal of Cognitive Education and Psychology*, *13*(2), 245–257. https://doi.org/10.1891/1945-8959.13.2.245

Chapter 7
Experiment in Education

7.1 Introduction

Understanding the many factors influencing education can be challenging because of the many complexities, and their interactions. Designing and implementing a quality education system require careful consideration of many factors, making education a "hot topic" for research. For this reason, investigations on education have become far more common in today's scientific literature (Wanti et al. 2022; Carvalho et al. 2023; Buchholtz et al. 2021, to mention a few).

One continuing problem is the different types of barriers which prevent groups from fully participating in the process of being educated. Issues involved with these barriers mean ensuring fair and transparent admissions processes, providing sufficient financial support for low-income students, and offering academic and social support services for student success. Creating a diverse and inclusive campus environment is also important to make all students feel welcome and valued, irrespective of their backgrounds.

When all is said and done, the need is for a holistic approach, which promotes equity (Prodan et al. 2015). At the level of specific action, this holistic approach must actively engage with communities hitherto underrepresented, with the goal to take action, which remedies problems of the past and prevents new problems in the future. The actions mean actively engaging with underrepresented communities, listening to their needs, and then doing specific actions rather than just talking. The actions may call for the involvement of community organizations and local leaders. In the end, equity in the access to higher education institutions (HEIs) is the goal toward, which the world of education must strive (Amaral 2022; Jurado de los Santos et al. 2020).

Higher education institutions play a vital role in shaping the future of students and society. The quality of teaching and learning is crucial in order for students to receive an education, which will prepare them for success. Educators ought to be

P. Papajorgji, H. Moskowitz, *The Mind of Everyday*,
https://doi.org/10.1007/978-3-031-78078-3_7

knowledgeable, passionate, dedicated to providing students with the tools needed for thriving in academia, and afterward in one's profession. The institutions are responsible as well, providing a supportive, inclusive learning environment, one which promotes collaboration, critical thinking, and creativity (Vasconcelos and Calheiros 2022). By consistently striving for excellence in teaching and learning, higher education institutions can empower students to achieve their full potential and make significant contributions to the world. In addition, higher education institutions should offer a diverse range of academic programs and extracurricular activities, designed for the varied interests and talents of students. By providing opportunities for students to explore different disciplines and engage in hands-on learning experiences, institutions can help them discover their passions and develop a well-rounded skill set. A culture of innovation and entrepreneurship inspires students to think outside the box and pursue ambitious goals. Overall, by equipping students with the knowledge, skills, and mind-sets needed to navigate an ever-changing world, higher education institutions can make a difference, doing so by shaping the future of individuals and society (Chugh et al. 2023; Sokoli et al. 2021; Noben et al. 2022).

In recent years, educational technology has become increasingly prevalent in higher education institutions. From online learning platforms to virtual reality simulations, new tools are revolutionizing the way students learn and engage with course materials. As technology continues to advance, it is crucial for universities to stay up to date with the latest trends and incorporate them into their curriculum. By doing so, institutions can provide students with a more interactive and personalized learning experience, ultimately preparing them for success in the digital age.

The shift toward integrating educational technology in higher education is not without its challenges, however. Faculty members may need additional training to effectively utilize these tools in their teaching. There is always the worry that technology will "somehow" replace traditional teaching methods. However, when implemented thoughtfully and strategically, educational technology has the potential to enhance student learning outcomes and improve overall academic performance. As the landscape of higher education continues to evolve, educational technology in higher education institutions must be prioritized. Institutions must embrace the opportunities that technology offers in order to better prepare their students for an ever-faster evolving future (Chugh et al. 2023).

In order to ensure the quality of education being provided, higher education institutions should institute robust assessment and accountability measures. These include regularly evaluating student learning outcomes, faculty performance, and institutional effectiveness, at a bare minimum. Transparent reporting of assessment results and accountability measures will help build trust with students, parents, and the broader community. Ultimately, a strong culture of assessment and accountability is essential for the success and reputation of higher education institutions. This culture of assessment and accountability also helps institutions to identify areas for improvement, implement changes, and track progress over time. By continuing efforts to enhance their educational offerings and outcomes, institutions can demonstrate their commitment to excellence and student success. Assessment and accountability are essential in higher education institutions (Rawlsyk 2018; Dawson and McHenry 2022).

The Mind Genomics study presented here investigates the four following topics, based upon the fact that the topics are the most important for education:

1. Equity and access in higher education institutions.
2. Quality of teaching and learning in higher education institutions.
3. Educational technology in higher education institutions.
4. Assessment and accountability in higher education institutions.

7.2 The Mind Genomics Experiments

This chapter focuses on the issue of education from two angles; ideas provided by human beings, the other using suggestions made by an AI-based system, the suggestions emerging from the LLM (large language model), Chat GPT3.5. Both studies use the method of Mind Genomics (Moskowitz et al. 2006; Veljko Milutinovic and Salom 2016; Gofman 2012). We analyze the two studies entirely separately, and then look at a metric, the IDT (Index of Divergent Thought) to give us a sense about the nature of the AI contribution. Are there any obvious differences, and if so which approach is better, and in fact does it make a difference?

The first step in a Mind Genomics research project identifies the important issues (Todri et al. 2021; Ilollari et al. 2020). The four basic questions or silos are the same, and so the same silos will be used in the two studies. Only the answers to the questions, the elements in the silo, will vary. In one study with elements from AI, the LLM (large language model) CHAT GPT3.5 will provide the elements, viz., the answers to questions (OpenAI 2024). In the parallel study human researchers without help from AI will provide the elements. Both studies will undergo parallel analyses of precisely the same sort. The objective is to determine whether there is an advantage, disadvantage, or neither by using different sources of elements. In simple terms, does the source of the elements make any difference.

Both studies were undertaken using the same Mind Genomics platform, presented in Chap. 6. Both studies were designed using a 4x4 Mind Genomics model. Each study worked with 100 respondents, each respondent testing 24 vignettes, and thus each studied worked with a database of 2400 observations. The two studies used the same type of respondents, provided by Luc.id, Inc., an online panel provider specializing in the recruiting of respondents.

The respondents agreeing to participate were given the following orientation:

> *Education is essential to everyone, and issues such as Quality of Teaching and Learning, Equity and Access to education, the Technology used, and Assessment and Accountability are the topics of this study. We want your opinion to understand the education systems' issues and challenges.*

We begin with the human-sourced elements. Table 7.1 presents the actual raw material, viz., the questions (silos) and the answers (elements) for the study with human-generated elements. Table 7.2 presents the actual raw material for the study with AI-generated elements.

Table 7.1 Answers (elements) generated human researcher

Elements defined by human researchers	
Question 1: Equity and access in higher education institutions: Do you think students with different background have equal access to quality education?	
A1	I believe that students from different backgrounds have equal access to quality education
A2	I do not think students from different backgrounds have equal access to quality education
A3	The access to quality education for students from different backgrounds varies depending on several socioeconomic factors
A4	While there are still disparities, the situation is improving, and we are moving toward more equal access to quality education for students from different backgrounds
Question 2: Quality of teaching and learning in higher education institutions: Do you think the effectiveness of teaching methods, curriculum design, and assessment practices play a crucial role in student success?	
B1	The effectiveness of teaching methods, curriculum design, and assessment practices are indispensable in shaping student success
B2	Factors such as student motivation, home environment, and access to resources also play crucial roles in students' success
B3	The impact of teaching methods, curriculum design, and assessment practices on student success can vary widely depending on the context and the specific needs of students
B4	Factors such as parental involvement, socioeconomic status, and access to educational resources are determinant in students' success
Question 3: Educational technology in higher education institutions: Do you think integrating technology into education is important for students' success?	
C1	The integration of technology in education is crucial for students' success
C2	Technology can be beneficial in the education process, but its importance should not be overstated
C3	The impact of technology in education can vary depending on the specific context and how it is implemented
C4	Overreliance on technology may lead to concerns about screen time, privacy, and equity issues
Question 4: Assessment and accountability in higher education institutions: Do you think developing fair and reliable methods for assessing student learning and teacher performance is essential?	
D1	Developing fair and reliable assessment methods is essential for students' success
D2	Overemphasis on standardized testing may lead to teaching to the test and narrowing the curriculum
D3	The importance of assessment methods can vary depending on the educational context and goals
D4	Overvaluing assessment methods can create undue stress for students and teachers

Table 7.2 Answers (elements) generated by AI

Elements (Answers) defined by AI-based system	
Question 1: Equity and access in higher education institutions: Do you think students with different background have equal access to quality education?	
A1	The students with Down's syndrome as well as those with mental retardation, autistic, and slow learners are integrative part of HEI auditors
A2	Students pertaining to wealthy families and non as well those pertaining to vulnerable groups have same access in HEI
A3	The number of foreign students in HEIs is continuously increasing
A4	HEI motivates dialogue, negotiation, cultural, and religious tolerance in its study programs and research projects
Question 2: Quality of teaching and learning in higher education institutions: Do you think the effectiveness of teaching methods, curriculum design, and assessment practices play a crucial role in student success?	
B1	Professors implement teaching methods that captures students' attention and develop their interests
B2	HEI frequently trains the professors concerning innovative teaching methods
B3	HEI implement student-centered methods, project-based teaching, and collaborative learning in all programs
B4	Students interact with each other in the classroom for problem solving issues
Question 3: Educational technology in higher education institutions: Do you think integrating technology into education is important for students' success?	
C1	HEI motivates the students toward critical thinking through implementation of various teaching and learning technologies
C2	The implementation of innovative teaching and learning motivates students toward creativity
C3	The implementation of teaching and learning technologies improves learning outcomes
C4	The implementation of teaching technologies helps professors to understand students' needs better
Question 4: Assessment and accountability in higher education institutions: Do you think developing fair and reliable methods for assessing student learning and teacher performance is essential?	
D1	HEI assesses students based on practical works, research projects, and technical exams
D2	HEI uses traditional and technological students' assessment methods
D3	HEI encourages students to evaluate professors teaching methods through semestral questionnaires
D4	HEI evaluates professors didactic and scientific performance on a yearly basis

7.3 Two-Sided Scale, Patterns of Scale Usage

Mind Genomics uses a five-point Likert scale. The respondent reads the vignette, one vignette at a time, and rates the vignette on the scale. The scale is not the traditional five-point scale with a single direction (bottom = not interested vs top = looks interesting). Rather, the scale comprises two sets of responses; interest (looks

Table 7.3 Frequency of ratings assigned to vignettes. The table compares the two studies, elements created by humans versus elements created by AI

	Original scale points Numbers in body of table are percent of times the rating scale was assigned to a vignette					
	1	2	3	4	5	Total
Source						
Total	6	8	22	23	41	100
Elements created by humans	6	6	19	23	46	100
Elements created by AI	5	10	25	23	37	100

interesting vs not interested) and probably of happening (it could happen vs. won't happen). In this way, the respondent provides two pieces of information in the same rating; judgment (interest) and belief (happen).

1 = Not interested ... AND ....won't happen
2 = Not interested ... BUT .. it could happen
3 = Can't honestly say
4 = Looks interesting .. BUT...won't happen
5 = Looks interesting .. AND .. it could happen.

The first "morphological" analysis looks at the proportions of rating points assigned to the vignettes. The focus is on the difference in the ratings distributions when we compare the ratings assigned to elements chosen by AI versus those chosen by people. Once again, there is no information about the specific elements besides their sources. Table 7.3 shows a slight difference between the two sources of ideas, primarily for the rating of "5," with the elements developed from human-sourced ideas generating many more ratings of "5" (looks interesting and could happen).

7.4 Transforming the Data to Prepare It for OLS (Ordinary Least Squares) Regression

We now turn to relating the presence/absence of the 16 elements to the newly derived rating. In the actual analysis as discussed below, we focus only on the combination of ratings 5 and 4, which together cover the dimension of "interest." We are focusing only on those elements that the respondent feels "*are of interest*." Thus, in the set of ratings below, a simplifying transformation was performed to convert ratings of 5 and 4 (both for 'Looks interesting) to the value 100, and in turn converted ratings of 3,2,1 (can't honestly say, not interest) to 0. Since the analysis will use OLS regression, we add a vanishingly small number (<10–5) to the newly created variable. That variable is Interest, which begins as 100 or 0, and now is converted to sum of the 100 or 0 and this vanishing small number. The addition ensures that in the analysis, the new dependent variable (Interest) will always show some variability across and within ratings assigned by the respondent, even if the respondent were to assign all vignettes 5,4 (all transformed values are 100), or to assign all values 3,2,1 (all transformed values are 0).

The next level of analysis looks at the performance of the elements. A key variable is R54x, which looks at the performance of the elements. R54x corresponds to the union of the two scale points featuring "it looks interesting." The analysis to link the response to the elements is OLS regression. The setup of the study permits the researcher to create equations by regression since the underlying experimental design for each respondent ensures that the 16 elements are presented in a way that makes them statistically independent of each other. One further property of Mind Genomics is that the use of a so-called permuted experimental design allows the respondent to evaluate a totally unique set of combinations. The benefit is one need not "know the answer" ahead of time. Everyone can be totally ignorant. And yet the data emerge.

The analysis for the Total Panel for each study creates this simple equation:

$$R54x = k1A1 + k2A2 \ldots + k16A16$$

We will also create the foregoing equation for gender (males vs. females), and for mind-sets (Mind-sets 1 of 2, 2 of 2; 1 of 3, 2 of 3, 3 of 3).

What is very important in this analysis emerges from looking at the elements that generate "high coefficients." High is operationally defined as a coefficient of +21 or higher, when the OLS regression does not use an additive constant. When we compare the sizes of the coefficients for the Total Panel versus for the mind-sets, we find that the coefficient for an element in Total Panel usually somewhere in the middle, as should be the case with an average. We learn from that the reality that the Total Panel ends up combining people with different ideas. It is like water streams meeting each other, either head on or at an angle, with the results that the streams cancel each other.

We can get a sense of the reality of cancellation by considering gender differences. Table 7.4 shows coefficients emerging from the human-based experiment. Table 7.5 shows coefficients emerging from the AI-based experiments. In both studies the coefficients for the Total Panel are low, and the coefficients for the genders are not particular higher either.

The samples of respondents for both studies were set up to present a distribution of age, education, gender, and residence. We focus here on gender differences.

Figure 7.1 gives a sense of the total sum of the four coefficients for each study (human-generated elements vs. AI-generated elements) and for each of the four silos. The sum of coefficients appears on the ordinate; the silo appears on the abscissa. The elements developed by the human researchers generated higher sums. This is a noteworthy finding. For today's level of sophisticated of LLM's such as Chat3.5 GPT, the human appears to outperform the machine, perhaps because the human researcher inserts a subtle "human sensitivity" into the elements not yet done by AI.

Table 7.4 Total and gender-based coefficients for human-generated elements

	Coefficients – Human Generated Elements (Transformed Dependent Variable = R54x (Interested))	**Total**		**Male**	**Female**
	Base Size	100		42	58
	Question A: Equity and Access: Do you think students with different background have equal access to quality education?				
A1	I believe that students from different backgrounds have equal access to quality education	17		16	18
A2	I do not think students from different backgrounds have equal access to quality education	17		16	18
A3	The access to quality education for students from different backgrounds varies depending on several socioeconomic factors	18		16	20
A4	While there are still disparities, the situation is improving, and we are moving towards more equal access to quality education for students from different backgrounds	18		19	17
	Question B: Teaching Quality: Do you think teaching methods improve teaching quality?				
B1	The effectiveness of teaching methods, curriculum design, and assessment practices are indispensable in shaping student success	18		15	20
B2	Factors such as student motivation, home environment, and access to resources also play crucial roles in students' success	19		22	17
B3	The impact of teaching methods, curriculum design, and assessment practices on student success can vary widely depending on the context and the specific needs of students	15		15	15
B4	Factors such as parental involvement, socioeconomic status, and access to educational resources are determinant in students' success	21		22	20
	Question C: Education Technology: Do you think technology enhances learning experiences?				
C1	The integration of technology in education is crucial for students' success	18		19	17
C2	Technology can be beneficial in the education process, but its importance should not be overstated	17		22	14
C3	The impact of technology in education can vary depending on the specific context and how it is implemented	18		18	17
C4	Overreliance on technology may lead to concerns about screen time, privacy, and equity issues	18		23	15
	Question D: Assessment and Accountability: Do you think developing reliable assessing methods is essential for quality teaching?				
D1	Developing fair and reliable assessment methods is essential for students' success	18		20	17
D2	Overemphasis on standardized testing may lead to teaching to the test and narrowing the curriculum	15		18	13
D3	The importance of assessment methods can vary depending on the educational context and goals	17		20	15
D4	Overvaluing assessment methods can create undue stress for students and teachers	18		19	17

Table 7.5 Total and gender-based coefficients for AI-generated elements

	AI-Generated Elements (Transformed Dependent Variable = R54x (Interested))	Total		Male	Female
	Base Size	100		47	53
	Question A: Equity and Access in Higher Education Institutions				
A1	Students with Down's syndrome, and those with mental retardation, autistic and slow learners, are integral parts of HEI auditors	23		26	21
A2	Students of wealthy or not wealthy families and those of vulnerable groups have the same access to HEI	22		25	19
A3	The number of international students in HEIs is continuously increasing	24		30	19
A4	HEI motivates dialogue, negotiation, cultural and religious tolerance in its study programs and research projects	25		30	20
	Question B: Quality of Teaching and Learning in Higher Education Institutions				
B1	Professors implement teaching methods that capture students' attention and develop their interests	21		21	21
B2	HEI frequently trains professors concerning innovative teaching methods	19		16	20
B3	HEI implements student-centered methods, project-based teaching, and collaborative learning in all programs	18		21	15
B4	Students interact with each other in the classroom for problem-solving issues	20		21	18
	Question C: Educational Technology in Higher Education Institutions				
C1	HEI motivates the students toward critical thinking through implementation of various teaching and learning technologies	16		21	12
C2	The implementation of innovative teaching and learning motivates students to creativity	20		20	21
C3	The implementation of teaching and learning technologies improves learning outcomes	18		20	16
C4	The implementation of teaching technologies helps professors to understand students' needs better	16		16	17
	Question D: Assessment and Accountability in Higher Education Institutions				
D1	HEI assesses students based on practical works, research projects, and technical exams	18		22	15
D2	HEI uses traditional and technological student assessment methods	22		24	20
D3	HEI encourages students to evaluate professors' teaching methods through semestral questionnaires	21		24	19
D4	HEI evaluates professors' didactic and scientific performance every year	19		25	14

7.5 AI Summarization of Strong Performing Elements for the Total Panel

As part of the Mind Genomics platform, an AI-based summarizer using Chap GPT3.5 (OpenAI 2024) explored different aspects of strong performing elements, viz., those operationally defined by scoring 21 or higher. The summarizer considers only those specific elements, and is done automatically, based upon prompts standard for the Mind Genomics platform. Table 7.6 shows the AI-summarization of elements generated by people. Table 7.7 shows the AI-summarization of elements generated by AI.

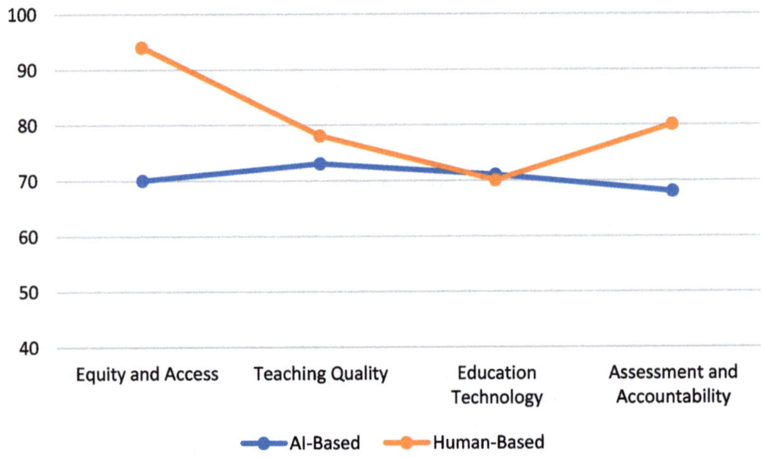

Fig. 7.1 Sums of the four coefficients in each silo, in each of the two experiments (human-generated elements; AI-generated elements)

Table 7.6 AI-summarization of strong performing elements generated by humans (Total Panel)

Idea summarization—Total panel—Human-generated ideas (coefficient > = 21) Interested in: Factors such as parental involvement, socioeconomic status, and access to educational resources are determinant in students' success
Create a label for this segment: Factors affecting student success
Describe this segment: This segment of the research study focuses on various factors that contribute to students' success, particularly the influence of parental involvement, socioeconomic status, and access to educational resources. Researchers are interested in understanding how these specific attributes impact students' academic achievements and overall educational outcomes
Parental involvement refers to the level of engagement and support provided by parents or caregivers in their child's education. This can include activities such as helping with homework, attending parent–teacher meetings, and participating in school activities. Researchers aim to examine whether higher levels of parental involvement lead to improved academic performance and increased motivation in students
Socioeconomic status (SES) refers to an individual or family's economic and social standing within society. It encompasses factors such as income, education, and occupation. In this segment, the researchers are interested in investigating how different socioeconomic backgrounds can affect students' educational outcomes. They might explore whether students from lower SES backgrounds face additional challenges in accessing quality educational resources and whether this impacts their academic success
Access to educational resources refers to the ability of students to utilize various materials, tools, and opportunities that enhance their learning experience. This can include access to textbooks, technology, extracurricular activities, advanced courses, and well-equipped educational facilities. The researchers seek to understand whether students with greater access to educational resources have higher chances of achieving academic success compared to those with limited access

(continued)

Table 7.6 (continued)

Overall, this segment of the research study aims to identify and analyze the influence of parental involvement, socioeconomic status, and access to educational resources on students' success. By exploring these factors, researchers hope to gain insights that can aid in developing strategies to improve educational outcomes and reduce disparities among different student demographics
Describe the attractiveness of this segment as a target audience: This segment can be considered highly attractive as a target audience. Factors such as parental involvement, socioeconomic status, and access to educational resources are well-established determinants of students' success. This implies that the segment includes individuals who are actively engaged in their child's education, have a higher socioeconomic status, and have the means to provide their children with adequate educational resources
The involvement of parents indicates a higher level of commitment toward their child's academic growth and development. Such parents are likely to be supportive and motivated to ensure their children's success in education. Therefore, targeting this segment can provide an opportunity to collaborate and engage with parents who are willing to actively participate in educational activities, programs, or initiatives
Moreover, the higher socioeconomic status of this segment suggests that they may have more financial resources, which can lead to better access to quality educational facilities, tutors, technology, and extracurricular activities. This increases the likelihood of their children receiving a well-rounded education and having access to a variety of opportunities, which can contribute to their overall success
Lastly, access to educational resources implies that this segment has the means to provide their children with books, educational materials, online resources, and other tools that can enhance learning. This further indicates their willingness to invest in their child's educational journey
Targeting this segment can offer a potentially receptive audience that values education and is actively involved in their child's academic development. Engaging with and providing relevant offerings, support, or resources to this segment can be seen as an investment in their children's success and can yield positive outcomes
List what is missing or should be known about this segment, in question form: 1. To what extent does parental involvement affect students' success? 2. What specific forms of parental involvement are most beneficial for students' success? 3. How does socioeconomic status influence students' success? 4. Are there specific elements of socioeconomic status, such as income level or parental occupation, that have a stronger impact on students' success? 5. How does access to educational resources impact students' success? 6. What specific educational resources (e.g., textbooks, technology, tutoring) are most crucial for students' success? 7. Is there a correlation between the level of parental involvement and access to educational resources? 8. Are there any other factors besides parental involvement, socioeconomic status, and access to educational resources that also contribute to students' success? 9. How do students' individual characteristics, such as motivation and learning abilities, interact with these external factors in determining their success? 10. Are there any cultural or community-specific factors that might influence the relationship between these attributes and students' success? 11. How do these attributes vary across different grade levels or educational stages? These are just a few questions that prompt to consider, and more specific questions could be formulated based on the context and objectives of the research study

(continued)

Table 7.6 (continued)

List and briefly describe attractive new or innovative products, services, experiences, or policies for this segment:
1. Virtual parent–school engagement platform: A user-friendly online platform that allows parents to actively participate in their child's education. It offers features such as real-time progress tracking, communication with teachers, access to educational resources, and interactive workshops
2. Socioeconomic empowerment scholarships: A scholarship program that aims to bridge the educational gap between students from different socioeconomic backgrounds. It provides financial support, mentorship, and access to additional educational resources to ensure that all students have equal opportunities to succeed
3. Mobile educational resource libraries: Mobile libraries equipped with educational resources, including books, tablets, and interactive learning tools that visit underprivileged areas regularly. This initiative aims to provide equal access to educational materials for students who may not have access to traditional libraries or resources at home
4. Digital learning platforms for low-income students: Online platforms that offer tailored educational content specifically designed for students from lower socioeconomic backgrounds. These platforms provide engaging and interactive learning experiences, helping students overcome barriers related to limited access to traditional educational resources
5. Affordable internet access: Partnerships with internet service providers and government initiatives to offer low-income families affordable internet access. This ensures that all students have equal opportunities to access online educational resources and participate in digital learning
6. Community learning centers: Establishing community learning centers in underprivileged areas, equipped with resources such as computers, books, and knowledgeable tutors. These centers provide a safe and supportive environment for students to study and access educational resources outside of school
7. Job placement programs and internships: Collaborations between schools and local businesses to provide job placement programs and internships for students from low-income backgrounds. These programs offer valuable hands-on experience, mentoring, and connections that can enhance students' employability and future success
8. Targeted financial aid programs: Developing financial aid programs specifically targeted at students whose parents have limited financial resources. These programs can support students through scholarships, grants, or low-interest loans, easing the burden of education costs and encouraging continued academic success
9. Student mentorship programs: Establishing mentorship programs where successful students act as mentors for their peers. This provides additional support, guidance, and motivation for students from disadvantaged backgrounds, helping them navigate the educational system and achieve their full potential
10. Enhancing teacher training: Providing comprehensive teacher training programs that focus on addressing the diverse needs of students from different socioeconomic backgrounds. These programs equip teachers with strategies and skills to effectively engage and support students facing various challenges, fostering a positive and inclusive learning environment

Table 7.8 shows the two and three mind-set coefficients for the elements emerging from human-generated elements (Mucherino et al. 2009). Table 7.9 shows the two and three mind-set coefficients for the elements emerging from the AI-generated elements. Once again, the actual equation is: INTERESTING (Binary, R54x) = k1(A1) + k2(A2) …k16(D4). The coefficients in Tables 7.6 and 7.7 show the results of six different OLS regressions for each study. The first OLS regression uses the 2400 rows of data, because all respondents are in the total panel. The remaining OLS regressions use only the sets of 24 rows of data for respondents

Table 7.7 AI-summarization of strong performing elements generated by AI (Total Panel)

Table 7.5 AI-summarization of elements generated by AI (coefficient for total panel > = 21)
Create a label for this segment: Inclusive and diverse student community
Describe this segment:
This segment refers to the attributes or characteristics of the total sample in a research study. It mentions that students with Down's syndrome, mental retardation, autism, and slow learners are considered integral parts of HEI auditors, suggesting that these students are included and actively involved in higher education institutions
Furthermore, it states that students from both wealthy and not wealthy families, as well as those belonging to vulnerable groups, have equal access to HEIs, emphasizing the importance of equal opportunities for all students regardless of their socioeconomic status or background
The segment also highlights the increasing number of international students attending HEIs, indicating a growing diversity and internationalization within higher education
Cultural and religious tolerance, as well as dialogue and negotiation, are promoted in HEI study programs and research projects, showcasing the institution's commitment to fostering inclusivity and respect for different cultural and religious backgrounds
Additionally, it mentions that professors employ teaching methods that engage students and cultivate their interests, suggesting a student-centered approach to education
Both traditional and technological student assessment methods are utilized by HEIs, indicating a blending of traditional educational practices with emerging technological advancements
Lastly, HEIs encourage students to provide feedback on professors' teaching methods through semestral questionnaires, demonstrating a commitment to continuous improvement and incorporating student input into the teaching and learning process
Overall, this segment highlights various aspects of the total sample in relation to student diversity, inclusivity, teaching methods, assessment practices, and student feedback within HEIs
Describe the attractiveness of this segment as a target audience: The segment of students with Down's syndrome, mental retardation, autism, and slow learners would be considered an important and inclusive target audience. By including these students, HEIs demonstrate a commitment to diversity and equal opportunities for education
Additionally, the fact that students from wealthy and vulnerable groups have the same access to HEIs indicates fairness and social equality within the institution. This can be attractive to potential students who value inclusivity and equal opportunities
The continuous increase in international students within HEIs suggests a vibrant and diverse community. This can be appealing to students who value multiculturalism and global perspectives in their education
The emphasis on dialogue, negotiation, cultural, and religious tolerance in the study programs and research projects of HEIs indicates a commitment to promoting understanding and acceptance. This can be attractive to individuals who value diversity, open-mindedness, and a supportive learning environment
The use of teaching methods that capture students' attention and develop their interests, as well as the use of both traditional and technological student assessment methods, suggests a modern and innovative approach to education. This can be appealing to students who seek dynamic and engaging learning experiences
Finally, the encouragement of student evaluation of professors' teaching methods through questionnaires demonstrates a willingness to listen and respond to student feedback. This signals a commitment to continuous improvement and student satisfaction, which can be attractive to prospective students

(continued)

Table 7.7 (continued)

Overall, the described attributes make this segment an attractive target audience, appealing to individuals who value inclusivity, equal opportunities, diversity, open-mindedness, innovation, and a supportive learning environment
List what is missing or should be known about this segment, in question form: 1. What specific accommodations or support services are provided for students with Down's syndrome, mental retardation, autism, and slow learners? 2. How does HEI ensure that students from vulnerable groups have the same access to higher education? 3. What efforts does HEI make to attract and support international students? 4. How does HEI promote dialogue, negotiation, cultural and religious tolerance in its study programs and research projects? 5. What specific teaching methods do professors use to capture students' attention and develop their interests? 6. What traditional and technological student assessment methods are utilized by HEI? 7. How does HEI encourage students to provide feedback on professors' teaching methods through semestral questionnaires? 8. What resources or support systems are available within HEI to assist students with their evaluations of professors' teaching methods? 9. How does HEI ensure that students' evaluations of professors are taken into consideration for improvement purposes? 10. Are there any specific programs or initiatives in place to enhance the overall experience and outcomes for students with disabilities or special needs within HEI?
List and briefly describe attractive new or innovative products, services, experiences, or policies for this segment: 1. Inclusive education programs: Develop new and innovative programs within higher education institutions (HEIs) that specifically cater to students with Down's syndrome, mental retardation, autism, and slow learners. These programs should focus on providing specialized support and accommodations to ensure the successful integration and learning of these students 2. Financial aid and scholarships: Introduce new financial aid programs and scholarships specifically targeted toward students from vulnerable groups. This would help ensure that students from economically disadvantaged backgrounds have the same access to HEIs as their wealthier counterparts 3. International student support services: Create comprehensive support services tailored toward international students. This could include language assistance programs, cultural orientation workshops, and mentoring programs to facilitate their integration into HEIs and the local community 4. Intercultural competence training: Implement mandatory intercultural competence training for all students and faculty members. This would promote cultural and religious tolerance, dialogue, and negotiation within HEIs, fostering a diverse and inclusive learning environment 5. Student engagement initiatives: Develop new teaching methods and strategies that aim to capture student attention and stimulate their interests. This could include interactive lectures, hands-on learning experiences, and use of technology and multimedia in the classroom 6. Innovative assessment methods: Incorporate innovative assessment methods that combine both traditional and technological approaches. This could involve the use of online quizzes, virtual simulations, and digital portfolios to enhance student learning and skills evaluation 7. Transparent feedback mechanisms: Establish a transparent system for students to evaluate professors' teaching methods, utilizing semester questionnaires or online platforms. This feedback can be used to continuously improve teaching practices and enhance the overall learning experience at HEIs

(continued)

Table 7.7 (continued)

Which messages will interest this segment?
Based on the given attributes, the following messages are likely to interest this segment:
1. Inclusive education: Highlight the fact that students with special needs such as Down's syndrome, mental retardation, autism, and slow learners are considered integral parts of the higher education institution (HEI) auditor program. Emphasize the inclusive and supportive environment provided for these students
2. Equal access for all: Emphasize that students from both wealthy and not wealthy families, as well as those from vulnerable groups, have equal access to HEIs. Promote the idea that financial status or background does not hinder educational opportunities
3. Increasing international diversity: Highlight the continuous increase in the number of international students in HEIs. Emphasize the benefits of a multicultural and diverse environment, including exposure to different perspectives and experiences
4. Promotion of dialogue and tolerance: Highlight how HEIs motivates dialogue, negotiation, cultural understanding, religious tolerance, and respect for diversity within study programs and research projects. Emphasize the importance of open-mindedness, collaboration, and mutual respect
5. Engaging teaching methods: Showcase that professors at HEIs implement teaching methods that capture students' attention and cater to their interests. Highlight innovative and interactive approaches to learning that help students actively engage and develop their skills
6. Varied assessment methods: Emphasize that HEIs use both traditional and technological student assessment methods. Highlight a balanced approach that incorporates different evaluation techniques and ensures fairness and accuracy
7. Student feedback: Highlight that HEIs actively encourage students to provide feedback on professors' teaching methods through semestral questionnaires. Emphasize that student voices are valued and taken into consideration for continuous improvement
Overall, messages that highlight inclusivity, equal access, diversity, engaging teaching methods, varied assessments, and student feedback are likely to resonate with this segment

falling into each mind-set. The result is fewer rows or data for the regression analysis for each mind-set.

Table 7.9 shows the two and three mind-set coefficients for the AI-generated elements.

Once again, the patterns generated by the strong-performing elements are reanalyzed by AI, just as was done for coefficients emerging from the study of human-generated elements, and for coefficients emerging from the study of AI-generated elements. This time, we have five key subgroups from which to choose, viz., mind-sets 1 and 2 of two, mind-sets 1,2,3 of three. For purposes of illustration, we consider only the former, the two mind-set solution. The richness of the results emerges when we look at the AI-summarization of strong-performing elements generated by humans (Table 7.10) and by AI (Table 7.11), respectively.

7.6 Creating Additional Knowledge by Transforming the Likert scale in Different Ways

Thus far we have only considered one transformation, viz., Interest', transforming ratings of 5 and 4 as 100, and ratings of 3, 2,1 as 0, respectively. Let us look at the range of transformations possible, viz., those transformations which "make sense."

Table 7.8 Two and three mind-set coefficients for human-generated elements

	Coefficients for the Human-Generated Elements (Dependent Variable = 54X, Interesting)	**Total**		**Mind-set 1 of 2**	**Mind-set 2 of 2**		**Mind-set 1 of 3**	**Mind-set 2 of 3**	**Mind-set 3 of 3**
	Base Size	100		55	45		39	31	30
	Question A: Equity and Access: Do you think students with different background have equal access to quality education?								
A1	I believe that students from different backgrounds have equal access to quality education	17		15	20		12	15	26
A2	I do not think students from different backgrounds have equal access to quality education	17		15	19		18	4	29
A3	The access to quality education for students from different backgrounds varies depending on several socioeconomic factors	18		19	18		20	9	27
A4	While there are still disparities, the situation is improving, and we are moving towards more equal access to quality education for students from different backgrounds	18		19	16		22		30
	Question B: Teaching Quality: Do you think teaching methods improve teaching quality?								
B1	The effectiveness of teaching methods, curriculum design, and assessment practices are indispensable in shaping student success	18		22	13		20	21	12
B2	Factors such as student motivation, home environment, and access to resources also play crucial roles in students' success	19		23	13		26	16	11
B3	The impact of teaching methods, curriculum design, and assessment practices on student success can vary widely depending on the context and the specific needs of students	15		17	12		18	15	8

B4	Factors such as parental involvement, socioeconomic status, and access to educational resources are determinant in students' success	21		24	17		23	19	17
	Question C: Education Technology: Do you think technology enhances learning experiences?								
C1	The integration of technology in education is crucial for students' success	18		14	22		10	26	19
C2	Technology can be beneficial in the education process, but its importance should not be overstated	17		11	24		9	35	11
C3	The impact of technology in education can vary depending on the specific context and how it is implemented	18		14	22		11	30	13
C4	Overreliance on technology may lead to concerns about screen time, privacy, and equity issues	18		12	26		9	33	15
	Question D: Assessment and Accountability: Do you think developing reliable assessing methods is essential for quality teaching?								
D1	Developing fair and reliable assessment methods is essential for students' success	18		25	10		24	19	11
D2	Overemphasis on standardized testing may lead to teaching to the test and narrowing the curriculum	15		19	9		20	16	9
D3	The importance of assessment methods can vary depending on the educational context and goals	17		26	7		27	19	6
D4	Overvaluing assessment methods can create undue stress for students and teachers	18		24	11		25	15	13

Tab. 7.8 (continued)

For each transformation we end up with 0's and 100's. We can take an average of these 0's and 100's for each meaningful transformation, and then compare the averages for the study with human-generated elements versus AI-generated elements. Table 7.12 shows the original scale values, their binary transformations, and the averages across all 2400 vignettes evaluated in each of the two studies. The averages give us an idea of how much better the human-generated portions perform. The two most revealing averages are R52x, which represents "could happen," and R54x,

Table 7.9 Two and three mind-set coefficients for the AI-generated elements

	Coefficients for the AI-Generated Elements (Dependent Variable = 54X, Interesting)	Total		Mind-set 1 of 2	Mind-set 2 of 2		Mind-set 1 of 3	Mind-set 2 of 3	Mind-set 3 of 3
	Base Size	100		47	53		31	40	29
	Question A: Equity and Access in Higher Education Institutions								
A1	Students with Down's syndrome, and those with mental retardation, autistic and slow learners, are integral parts of HEI auditors	23		25	21		20	19	31
A2	Students of wealthy or not wealthy families and those of vulnerable groups have the same access to HEI	22		24	20		18	17	32
A3	The number of international students in HEIs is continuously increasing	24		24	24		16	21	34
A4	HEI motivates dialogue, negotiation, cultural and religious tolerance in its study programs and research projects	25		22	28		19	25	31
	Question B: Quality of Teaching and Learning in Higher Education Institutions								
B1	Professors implement teaching methods that capture students' attention and develop their interests	21		20	22		18	21	26
B2	HEI frequently trains professors concerning innovative teaching methods	19		21	16		19	13	26
B3	HEI implements student-centered methods, project-based teaching, and collaborative learning in all programs	18		19	16		21	14	18
B4	Students interact with each other in the classroom for problem-solving issues	20		21	18		25	18	18

	Question C: Educational Technology in Higher Education Institutions								
C1	HEI motivates the students toward critical thinking through implementation of various teaching and learning technologies	16		18	14		30	13	6
C2	The implementation of innovative teaching and learning motivates students to creativity	20		23	19		28	19	15
C3	The implementation of teaching and learning technologies improves learning outcomes	18		20	15		28	16	8
C4	The implementation of teaching technologies helps professors to understand students' needs better	16		20	13		26	12	13
	Question D: Assessment and Accountability in Higher Education Institutions								
D1	HEI assesses students based on practical works, research projects, and technical exams	18		12	23		17	25	8
D2	HEI uses traditional and technological student assessment methods	22		16	28		14	30	19
D3	HEI encourages students to evaluate professors' teaching methods through semestral questionnaires	21		16	26		14	28	18
D4	HEI evaluates professors' didactic and scientific performance every year	19		12	25		15	24	17

Tab. 7.9 (continued)

which represents "looks interesting." When evaluating both BDVs, we found that the human-created elements performed better than the AI-created elements.

The BimiLeap program records the number of 100ths of a second between the appearance of the vignette on the respondent's screen and the response. To accommodate multitasking, the program transforms any response time of 9 s or more to 9 s. The average response time for elements generated is 3.2 s for vignettes created by humans, and 3.5 s for vignettes created by AI.

Continuing the comparison, we can do the OLS regression with these newly created variables to compare the patterns generated for the two studies. The data have already been prepared by the prophylactic addition of the vanishingly small random

Table 7.10 AI summarization of strong performing elements for the two mind-set solution emerging from elements generated by humans

Elements generated by humans (coefficient > =21) **Mind-set 1 of 2—Top** The following is a description of attributes from mind-Set 1 of 2 Interested in: The effectiveness of teaching methods, curriculum design, and assessment practices are indispensable in shaping student success; factors such as student motivation, home environment, and access to resources also play crucial roles in students' success; factors such as parental involvement, socioeconomic status, and access to educational resources are determinant in students' success; developing fair and reliable assessment methods is essential for students' success; the importance of assessment methods can vary depending on the educational context and goals; overvaluing assessment methods can create undue stress for students and teachers;
Create a label for this segment: Factors influencing student success and the importance of effective teaching and assessment methods
Describe this segment: This segment describes the various factors and elements that contribute to student success in a research study. It highlights the significance of effective teaching methods, curriculum design, and assessment practices in shaping student achievement. Additionally, it acknowledges the impact of factors like student motivation, home environment, and access to resources in determining student success. The segment also emphasizes the role of parental involvement, socioeconomic status, and access to educational resources as determinants of student success. It stresses the importance of developing fair and reliable assessment methods for ensuring student success. Furthermore, it notes that the significance of assessment methods may vary depending on the educational context and goals. Lastly, the segment warns against overvaluing assessment methods, as it can result in undue stress for both students and teachers
Describe the attractiveness of this segment as a target audience: This segment as a target audience would be highly attractive for those involved in the education field. The description highlights the importance of teaching methods, curriculum design, and assessment practices in shaping student success, which suggests that individuals who are interested in improving educational outcomes would find value in this information. Additionally, the mention of factors such as student motivation, home environment, and access to resources emphasizes the holistic perspective required for understanding and addressing student success, making it appealing to educators, policymakers, and researchers aiming to create inclusive and effective educational environments. The reference to parental involvement, socioeconomic status, and access to educational resources further emphasizes the broader societal factors impacting student success, making it relevant for individuals involved in educational equity and social justice initiatives. The segment also acknowledges the need for fair and reliable assessment methods, which would be appealing to educators and administrators looking for guidance in improving their assessment practices. However, the segment also cautions against overvaluing assessment methods, which could resonate with individuals seeking a more balanced and healthy approach to evaluating student success. Overall, the segment's focus on various aspects of education and student success, along with its acknowledgment of both strengths and areas for improvement, makes it an attractive target audience for researchers, educators, policymakers, and those interested in improving educational outcomes

(continued)

Table 7.10 (continued)

List what is missing or should be known about this segment, in question form: 1. What specific teaching methods were used in the research study? 2. How was the curriculum designed in the research study? 3. What assessment practices were implemented in the research study? 4. How was student motivation measured in the research study? 5. How was the home environment of students taken into account in the research study? 6. What educational resources were considered in the research study? 7. How was parental involvement measured in the research study? 8. How was socioeconomic status of students determined in the research study? 9. What educational resources were made accessible to students in the research study? 10. How were fair and reliable assessment methods developed in the research study? 11. What specific educational context and goals were considered in the research study? 12. How was the level of stress experienced by students and teachers measured in the research study? 13. Were there any specific challenges or limitations faced in implementing the assessment methods in the research study?

(continued)

Table 7.10 (continued)

List and briefly describe attractive new or innovative products, services, experiences, or policies for this segment: 1. Personalized learning platforms: These platforms utilize adaptive technology to tailor the learning experience to individual student needs, allowing for personalized curriculum and teaching methods. They provide real-time feedback and support to enhance student success 2. Collaborative learning environments: These environments promote cooperative and interactive learning experiences. They utilize technology to facilitate group work, encourage peer-to-peer learning, and foster a sense of responsibility and engagement among students 3. Home–school partnerships: Establishing strong connections between home and school is crucial for student success. Innovative partnership programs can include regular communication between teachers and parents, workshops, and resources for parents to support their child's learning at home, and involvement of parents in school decision-making processes 4. Technology integration: Integrating technology into educational practices can enhance student success. Innovative products and tools like interactive learning apps, virtual reality experiences, and multimedia resources can make learning more engaging, accessible, and relevant for students 5. Socioeconomic support programs: Recognizing the impact of socioeconomic factors on student success, implementing comprehensive support programs can make a significant difference. This may include providing financial assistance for educational resources, access to tutoring or mentoring programs, and offering additional educational opportunities for disadvantaged students 6. Alternative assessment methods: Moving away from traditional assessment methods, innovative approaches like project-based assessments, portfolio evaluations, and competency-based assessments can offer a more comprehensive understanding of student skills and knowledge. These methods focus on practical application and skill development instead of relying solely on standardized tests 7. Mental health and wellness initiatives: Recognizing the importance of mental health in student success, implementing programs that foster Well-being and resilience can enhance learning outcomes. This could include providing mindfulness training, counseling services, and promoting a positive school environment that prioritizes student Well-being 8. Open educational resources: Access to educational resources can be a determinant of student success. Open educational resources, including freely available textbooks, online courses, and learning materials, can bridge the gap and provide equitable access to quality resources for all students 9. Flexible learning environments: Offering flexible learning environments that cater to students' individual needs and learning preferences can boost student success. This may involve providing options for online or blended learning, flexible scheduling, and diversified classroom setups to accommodate different learning styles 10. Teacher professional development: Supporting teachers' ongoing professional development can enhance their effectiveness in improving student success. Innovative programs that provide opportunities for personalized training, collaboration, and access to cutting-edge teaching methods and resources can empower educators to make a positive impact on student outcomes

(continued)

Table 7.10 (continued)

Which messages will interest this segment? This segment will likely be interested in the following messages: 1. The effectiveness of teaching methods, curriculum design, and assessment practices are crucial in shaping student success 2. Factors such as student motivation, home environment, and access to resources also play significant roles in students' success 3. Factors such as parental involvement, socioeconomic status, and access to educational resources are determinants in students' success 4. Developing fair and reliable assessment methods is essential for students' success 5. The importance of assessment methods can vary depending on the educational context and goals 6. Overvaluing assessment methods can create undue stress for students and teachers
Mind-set 2 of 2—Top **Elements generated by humans** **(coefficients > =21)** Interested in: The integration of technology in education is crucial for students' success; technology can be beneficial in the education process, but its importance should not be overstated; the impact of technology in education can vary depending on the specific context and how it is implemented; overreliance on technology may lead to concerns about screen time, privacy, and equity issues;
Create a label for this segment: "Perspectives on the integration of Technology in Education"
Describe this segment: This segment of the research study has a mixed viewpoint regarding the integration of technology in education. While acknowledging the crucial role technology plays in students' success, there is also a cautionary approach emphasizing that technology's significance should not be exaggerated. The segment recognizes that the impact of technology in education can differ based on the specific context and the manner in which it is implemented. There is a concern about overreliance on technology, raising issues related to screen time, privacy, and equity
Describe the attractiveness of this segment as a target audience: The target audience described in this segment would be highly attractive for various reasons. Firstly, their belief in the integration of technology in education signals a positive attitude toward adopting technological advancements in the learning process. This aligns well with the current trend of digitization and the increasing use of technology in classrooms
Additionally, their understanding that technology should not be overstated in terms of its importance indicates a balanced perspective and a critical approach toward its implementation. This suggests that they would be open to exploring the benefits of technology while also considering its limitations and potential drawbacks
The recognition that the impact of technology in education can vary depending on the specific context and implementation showcases their awareness of the need for adaptability and customization in educational practices. This audience segment would likely value the importance of tailoring technology to suit the unique needs and requirements of different educational settings and student populations
Furthermore, their concerns about overreliance on technology highlight a cautious and mindful approach. This audience segment understands the potential issues related to excessive screen time, privacy, and equity, which are important considerations in the modern education landscape. Their concerns demonstrate a desire for a balanced approach that prioritizes student Well-being and fairness in educational opportunities

(continued)

Table 7.10 (continued)

Overall, this segment appears to be well-informed, open-minded, and receptive to technological advancements in education. They value the integration of technology while also being conscious of its potential drawbacks. As such, they would likely be an attractive target audience for educational technology providers, policymakers, and researchers seeking to cater to the diverse needs and concerns in the field of technology-integrated education
Explain why this segment might not be attractive as a target audience: This segment might not be attractive as a target audience because they hold mixed opinions about the integration of technology in education. While they acknowledge the importance and benefits of technology, they also highlight the need to not overstate its significance. This suggests that they may not be fully convinced or enthusiastic about the use of technology in education
Additionally, their emphasis on the varying impact of technology depending on the specific context and implementation indicates that they may have reservations about the effectiveness or appropriateness of technology in certain educational settings. This uncertainty could make it difficult to engage or persuade them as a target audience
Moreover, their concerns about overreliance on technology leading to issues such as excessive screen time, privacy concerns, and equity issues suggests a cautious attitude toward technology integration. They may be more inclined to prioritize traditional teaching methods or place greater emphasis on addressing the potential drawbacks of technology, which may not align with the goals or strategies of companies or organizations targeting technology integration in education
Overall, due to their mixed opinions, reservations, and concerns, this segment may not be receptive to marketing campaigns or initiatives that heavily promote or rely on the integration of technology in education
List what is missing or should be known about this segment, in question form: 1. How does the integration of technology in education specifically contribute to students' success? 2. What are the specific benefits of using technology in the education process? 3. How does the importance of technology in education compare to other factors that contribute to students' success? 4. What are the specific contextual factors that can affect the impact of technology in education? 5. How can technology be effectively implemented in different educational contexts? 6. What are the potential negative effects of overreliance on technology in education? 7. How does screen time affect students' learning outcomes? 8. What are the privacy concerns related to technology use in education? 9. What are the equity issues associated with technology integration in education? 10. Are there any specific guidelines or best practices for balancing technology use and maintaining equity in education?
List and briefly describe attractive new or innovative products, services, experiences, or policies for this segment:

(continued)

Table 7.10 (continued)

1. Personalized learning platforms: These platforms use technology to adapt educational content to each student's individual needs, providing a personalized learning experience. They offer customized lesson plans, assessments, and feedback, allowing students to progress at their own pace 2. Virtual reality (VR) educational experiences: VR technology allows students to explore virtual worlds and engage in immersive learning experiences. It can bring complex concepts to life and enhance understanding in subjects such as science, history, and geography. VR also enables students to visit inaccessible places, historical events, or even outer space 3. Online mentoring programs: These programs connect students with experienced mentors or professionals in their field of interest. Through video conferencing or online platforms, students can receive guidance, advice, and support to enhance their learning and career development 4. Gamification in education: Gamification integrates game elements such as competition, rewards, and achievements into educational activities. It helps to engage students, promote active learning, and foster a sense of accomplishment and motivation. Gamification can be applied to various subjects, making learning more enjoyable and interactive 5. Technology infrastructure for underserved schools: To address concerns about equity, innovative policies and initiatives should focus on providing adequate technology infrastructure to schools in disadvantaged areas. This includes ensuring high-speed internet access, providing devices such as laptops or tablets, and training teachers to effectively integrate technology into their teaching practices 6. Data privacy and security solutions: Given the concerns about privacy in technology integration, innovative products and services may aim to provide robust data protection measures for educational institutions. This can include encryption technologies, secure cloud storage, and compliance solutions to protect student data and privacy 7. Digital assessment tools: Traditional paper-and-pencil assessments are being replaced by digital assessment tools that offer more interactive and detailed evaluation methods. These tools can provide immediate feedback, measure progress over time, and offer adaptive assessment algorithms to gauge students' strengths and weaknesses accurately 8. Mobile learning apps: Mobile learning apps enable students to access educational content anytime and anywhere. These apps can offer interactive lessons, study materials, quizzes, and collaboration tools, promoting continuous learning and supporting students' engagement outside the classroom Overall, these innovative products, services, experiences, or policies aim to harness the benefits of integrating technology in education while also addressing concerns regarding privacy, screen time, and equity

(continued)

Table 7.10 (continued)

Which messages will interest this segment? This segment will be most interested in the following messages: 1. The integration of technology in education is crucial for students' success: This message aligns with their belief that technology is beneficial in education. Highlighting how technology can enhance student learning and improve academic outcomes will capture their attention 2. The impact of technology in education can vary depending on the specific context and implementation: This message acknowledges their understanding that technology's effectiveness is not universal. Emphasizing the importance of tailor-made approaches and considering the specific needs and goals of different educational contexts will resonate with them 3. Overreliance on technology may lead to concerns about screen time, privacy, and equity issues: This message addresses their concerns about potential negative consequences of technology overuse. Discussing strategies to mitigate these concerns, such as establishing balanced screen time guidelines, ensuring student privacy, and addressing equity gaps, will be of interest to this segment 4. Technology can be beneficial in the education process, but its importance should not be overstated: This message aligns with their viewpoint that while technology has its advantages, it should not be seen as a cure-all solution for all educational challenges. Highlighting the importance of a balanced approach that integrates technology with other instructional strategies will resonate with them Overall, this segment is interested in messages that emphasize the benefits of technology in education while addressing concerns about overreliance, privacy, equity, and the need for context-specific implementation

Table 7.11 AI summarization of strong performing elements for the two mind-set solution emerging from elements generated by humans

Elements generated by AI—Mind-sets **Mind-set 1 of 2—Top** **Elements generated by AI** **Coefficient > =21** Interested in: Students with Down's syndrome, and those with mental retardation, autistic, and slow learners, are integral parts of HEI auditors; students of wealthy or not wealthy families and those of vulnerable groups have the same access to HEI; the number of international students in HEIs is continuously increasing; HEI motivates dialogue, negotiation, cultural and religious tolerance in its study programs and research projects; HEI frequently trains professors concerning innovative teaching methods; students interact with each other in the classroom for problem-solving issues; the implementation of innovative teaching and learning motivates students to creativity;
Create a label for this segment: "Inclusive and diverse HEI Community focused on innovation and collaboration"
Describe this segment: This segment describes the inclusivity and diversity within higher education institutions (HEIs). It highlights that HEIs include students with Down's syndrome, mental retardation, autism, and slow learners. It also states that students from both wealthy and vulnerable backgrounds have equal access to HEIs
Furthermore, the segment mentions the increasing number of international students in HEIs, suggesting a global and multicultural environment. It emphasizes the importance of dialogue, negotiation, cultural, and religious tolerance in the study programs and research projects of HEIs

(continued)

Table 7.11 (continued)

The segment also mentions that HEIs frequently train professors in innovative teaching methods, suggesting a focus on modern and effective pedagogy. It highlights the importance of student interaction in the classroom for problem-solving purposes
Lastly, the segment states that the implementation of innovative teaching and learning methods within HEIs motivates students' creativity. This suggests an environment that fosters and encourages student innovation and originality
Describe the attractiveness of this segment as a target audience: This segment of students can be seen as an attractive target audience for several reasons
Firstly, including students with Down's syndrome, mental retardation, autism, and slow learners as integral parts of higher education institutions (HEIs) shows a commitment to diversity and inclusivity. This demonstrates that the HEIs value the unique abilities and perspectives that these students bring and are dedicated to providing equal opportunities for their education
Secondly, the equal access to HEIs for students from both wealthy and not wealthy families, as well as vulnerable groups, is appealing. This indicates that the HEIs aim to bridge the socioeconomic gap and ensure that education is accessible to all, regardless of their financial background or social status
Thirdly, the increasing number of international students in HEIs suggests that these institutions have a global outlook and value diversity in terms of cultural backgrounds and experiences. This can create a rich and diverse learning environment that fosters cross-cultural understanding and appreciation
Additionally, the emphasis on promoting dialogue, negotiation, cultural and religious tolerance in study programs and research projects demonstrates a commitment to fostering an inclusive and respectful learning environment. This can attract students who value open-mindedness, multiculturalism, and the opportunity to engage in meaningful discussions and interactions
The frequent training of professors concerning innovative teaching methods shows the HEIs' dedication to providing high-quality education and staying updated with the latest pedagogical approaches. This can be appealing to students who are seeking a dynamic and engaging learning experience
Lastly, the focus on students interacting with each other in the classroom for problem-solving fosters a collaborative and interactive learning environment. This can attract students who prefer hands-on learning and value the opportunity to work together with their peers to solve real-world problems
Overall, this segment of students can be considered attractive as a target audience due to the HEIs' commitment to diversity, inclusivity, equal access, global perspective, innovation in teaching, and interactive learning
List what is missing or should be known about this segment, in question form: 1.What is the specific number or percentage of students with Down's syndrome, mental retardation, autism, and slow learners in HEI auditors? 2. How does HEI ensure that students from wealthy and not wealthy families, as well as those from vulnerable groups, have equal access to HEIs? 3. How much has the number of international students in HEIs increased over time? What is the current proportion of international students compared to domestic students? 4. Can specific examples be provided to demonstrate how HEI motivates dialogue, negotiation, cultural, and religious tolerance in study programs and research projects? 5. How frequently are professors trained in innovative teaching methods by HEI? 6. How do students interact with each other in the classroom for problem-solving issues? Are there specific strategies or platforms used to facilitate this interaction? 7. Can specific examples be provided to illustrate how the implementation of innovative teaching and learning motivates students' creativity?

(continued)

Table 7.11 (continued)

List and briefly describe attractive new or innovative products, services, experiences, or policies for this segment: 1. Inclusive education programs: The development of specialized programs within higher education institutions (HEIs) that are specifically designed to cater to the needs of students with disabilities such as Down's syndrome, mental retardation, autism, and slow learners. These programs would provide the necessary support and accommodations to enable these students to fully participate in higher education 2. Financial aid programs: Innovative financial aid initiatives that aim to ensure equal access to HEIs for students from both wealthy and vulnerable groups. These programs could provide scholarships, grants, or other forms of financial support to bridge the economic gap and make higher education more accessible to all 3. International student support services: As the number of international students in HEIs continues to increase, there is a need for specialized support services to cater to their unique needs. Innovative programs can be developed to provide language and cultural support, as well as mentorship and integration programs, to help these students feel more included and supported within the HEI community 4. Diversity and inclusion training: HEIs can develop innovative training programs for professors, focusing on promoting cultural and religious tolerance, dialogue, and negotiation within the classroom. This kind of training would equip professors with the necessary skills to create inclusive and accepting classroom environments, where students from diverse backgrounds can freely express themselves and engage in meaningful discussions 5. Technology-enabled teaching methods: HEIs can explore innovative teaching methods that leverage technology to enhance learning and encourage creativity among students. This could include the use of interactive platforms, virtual reality (VR), or gamified approaches to foster creativity, critical thinking, and problem-solving skills among students 6. Collaborative learning initiatives: Creating opportunities for students to collaborate and interact with each other in the classroom for problem-solving issues can be an attractive innovation. Group projects, case studies, or cooperative learning activities can be incorporated into the curriculum to foster teamwork, communication skills, and a sense of community among students 7. Innovation hubs and maker spaces: HEIs can establish innovation hubs or maker spaces on campus, where students can explore and experiment with new ideas, technologies, and creative projects. These spaces would provide resources, mentorship, and interdisciplinary collaboration opportunities to foster innovation, entrepreneurship, and hands-on learning experiences Overall, these innovative products, services, experiences, or policies aim to create a more inclusive, diverse, and creative learning environment within HEIs, catering to the unique needs of different student segments and fostering their overall success

(continued)

Table 7.11 (continued)

Which messages will interest this segment? 1) Students with Down's syndrome, mental retardation, autism, and slow learners are integral parts of HEI auditors. This message highlights the inclusivity and support available for students with disabilities at the HEI 2) Students of wealthy or not wealthy families and those of vulnerable groups have the same access to HEI. This message emphasizes equal opportunities and accessibility for students from different socioeconomic backgrounds 3) The number of international students in HEIs is continuously increasing. This message highlights the diverse and multicultural environment of the HEI, which can be appealing to students interested in experiencing different cultures and perspectives 4) HEI motivates dialogue, negotiation, cultural and religious tolerance in its study programs and research projects. This message emphasizes the importance of fostering inclusivity, understanding, and respect for different cultures and religions, which may interest students who value diversity and open-mindedness 5) HEI frequently trains professors concerning innovative teaching methods. This message suggests that the HEI is committed to providing high-quality education by continuously improving teaching methods, which may interest students who are looking for a dynamic and engaging learning environment 6) Students interact with each other in the classroom for problem-solving issues. This message emphasizes collaborative learning and problem-solving skills, which may interest students who prefer interactive and participatory learning experiences 7) The implementation of innovative teaching and learning motivates students to creativity. This message suggests that the HEI values and promotes creativity and innovation in its teaching approach, which may interest students who are looking for opportunities to express their creative abilities
Mind-set-set 2 of 2—Top **Elements generated by AI** **Coefficients > = 21** Interested in: Students with Down's syndrome, and those with mental retardation, autistic, and slow learners, are integral parts of HEI auditors; the number of international students in HEIs is continuously increasing; HEI motivates dialogue, negotiation, cultural and religious tolerance in its study programs and research projects; professors implement teaching methods that capture students' attention and develop their interests; HEI assesses students based on practical works, research projects, and technical exams; HEI uses traditional and technological student assessment methods; HEI encourages students to evaluate professors' teaching methods through semestral questionnaires; HEI evaluates professors' didactic and scientific performance every year;
Create a label for this segment: "Inclusive education and teaching methods in HEIs"
Describe this segment: This segment consists of students with various disabilities such as Down's syndrome, mental retardation, autism, and slow learners who are enrolled in higher education institutions (HEIs). The HEIs also have an increasing number of international students. The HEIs emphasize promoting dialogue, negotiation, cultural and religious tolerance in their study programs and research projects. The professors in these institutions employ teaching methods that effectively engage students and develop their interests. The HEIs assess students based on practical works, research projects, and technical exams, utilizing both traditional and technological assessment methods. Additionally, the HEIs encourage students to evaluate the teaching methods of professors through semestral questionnaires, while also conducting yearly evaluations of professors' didactic and scientific performance

(continued)

Table 7.11 (continued)

Describe the attractiveness of this segment as a target audience: The segment described in the attributes is highly diverse and inclusive, as it includes students with Down's syndrome, mental retardation, autism, and slow learners. This demonstrates a commitment to providing equal opportunities and support to individuals with various learning challenges
Additionally, the increasing number of international students in higher education institutions (HEIs) suggests a global appeal and a recognition of the value of diversity and cultural exchange within the institution. This can enrich the learning experience and facilitate the development of a global perspective among students
The emphasis on dialogue, negotiation, cultural, and religious tolerance in study programs and research projects indicates a commitment to promoting inclusivity and understanding among students. This can create a welcoming and supportive environment for individuals from different backgrounds, fostering a sense of belonging and acceptance
The implementation of teaching methods that capture students' attention and develop their interests demonstrates a proactive approach to engaging students. This can enhance the learning experience and increase student motivation and satisfaction
The assessment methods employed by HEIs, including practical works, research projects, and technical exams, suggest a focus on practical skills development and critical thinking, which can be attractive to students seeking hands-on learning experiences
The use of both traditional and technological assessment methods acknowledges the importance of adapting to advancements in education technology while maintaining traditional evaluation approaches. This can cater to the preferences and needs of a diverse student body
HEI's encouragement of students to evaluate professors' teaching methods through semestral questionnaires demonstrates a commitment to continuous improvement and responsiveness to student feedback. This can indicate a student-centric approach and a willingness to address any concerns or areas for improvement
Finally, the annual evaluation of professors' didactic and scientific performance suggests a commitment to maintaining high-quality teaching standards. This can provide students with confidence in the expertise and competence of their professors
Overall, the segment described in the attributes appears attractive as a target audience due to its inclusive nature, commitment to diversity and cultural understanding, engaging teaching methods, varied assessment approaches, responsiveness to student feedback, and emphasis on high-quality teaching

(continued)

Table 7.11 (continued)

List what is missing or should be known about this segment, in question form: 1. How many students with Down's syndrome, mental retardation, autism, and slow learners are enrolled in HEIs? 2. What specific support services or accommodations are provided for students with disabilities in HEIs? 3. What is the percentage of international students in HEIs? 4. What factors contribute to the increasing number of international students in HEIs? 5. How does HEI specifically motivate dialogue, negotiation, cultural, and religious tolerance in study programs and research projects? 6. What specific teaching methods do professors use to capture students' attention and develop their interests? 7. How does HEI assess practical works, research projects, and technical exams? 8. What traditional student assessment methods are used by HEI? 9. What technological student assessment methods are used by HEI? 10. What feedback or data is collected from students regarding professors' teaching methods through semestral questionnaires? 11. How does HEI evaluate professors' didactic performance? 12. How does HEI evaluate professors' scientific performance? 13. Are there any additional support services or resources available to students in HEIs? 14. How does HEI ensure inclusivity and equity for students with disabilities, international students, and other diverse groups? 15. Are there any initiatives or programs in place to support cultural exchange and understanding among students in HEIs?
List and briefly describe attractive new or innovative products, services, experiences, or policies for this segment: 1. Inclusive learning programs: HEIs can develop specialized programs that cater to the unique needs of students with Down's syndrome, mental retardation, autism, and slow learners. These programs can provide additional support, adapted teaching methods, and individualized learning plans to ensure the success of these students in higher education 2. International student support services: HEIs can enhance their support services for international students by offering orientation programs, language support, cultural integration initiatives, and assistance in finding accommodation. This would help international students feel welcomed and supported during their stay at the institution 3. Multicultural education: HEIs can promote multicultural education by incorporating diverse perspectives, cultural awareness, and religious tolerance into their study programs and research projects. This could involve offering courses or workshops on intercultural communication, diversity management, and global citizenship 4. Engaging teaching methods: HEIs can encourage professors to use innovative teaching methods that not only capture students' attention but also foster their interests and promote active learning. This could include incorporating technology, interactive activities, group assignments, and real-world case studies into the curriculum 5. Diverse assessment methods: HEIs can adopt a combination of traditional and technological student assessment methods. This could involve using online quizzes, simulation exercises, practical assignments, and research projects to evaluate students' knowledge and skills. Ensuring a diverse range of assessment methods can better measure students' understanding and competence in their respective fields 6. Student feedback mechanisms: HEIs can establish a system for students to provide feedback on professors' teaching methods through regular questionnaires or surveys. This feedback can be utilized to improve teaching practices, address any concerns, and identify areas of improvement for the professors 7. Periodic performance evaluation: HEIs should continue evaluating professors' didactic and scientific performance on an annual basis. This process can help identify and recognize exceptional faculty members, provide constructive feedback to professors for professional development, and ensure the overall quality of teaching within the institution

(continued)

Table 7.11 (continued)

Which messages will interest this segment?
1. HEI values and includes students with Down's syndrome, mental retardation, autism, and slow learners, making them integral parts of the institution
2. The number of international students in HEIs is continuously increasing, indicating a diverse and multicultural environment
3. HEI actively promotes dialogue, negotiation, cultural understanding, and religious tolerance in its study programs and research projects
4. Professors at HEI implement teaching methods that capture students' attention and develop their interests
5. HEI assesses students based on practical works, research projects, and technical exams, providing a hands-on learning experience
6. HEI utilizes a combination of traditional and technological student assessment methods, ensuring a comprehensive evaluation process
7. HEI values student feedback and encourages them to evaluate professors' teaching methods through semestral questionnaires
8. HEI regularly evaluates professors' didactic and scientific performance, ensuring high teaching standards

Table 7.12 Transformations performed on the evaluation scale, and the comparison of average binary variables for human versus AI

	The original points on the rating scale						
	1	2	3	4	5		
Interesting	N	N	DK	Y	Y		
Happen	N	Y	DK	N	Y		
						Study elements generated by	
						Human	**AI**
R1x	100	0	0	0	0	7	5
R2x	0	100	0	0	0	6	10
R3x	0	0	100	0	0	19	25
R4x	0	0	0	100	0	23	23
R5x	0	0	0	0	100	46	37
R54x	0	0	0	100	100	69	60
R52x	0	100	0	0	100	52	47
R41x	100	0	0	100	0	13	15
R21x	100	100	0	0	0	29	28

numbers at the time of the transformation, preventing the OL regression from "crashing."

We employ OLS regression to relate the newly constructed BDVs and response time to the presence/absence of elements, which leads to a greater knowledge of the human-generated elements versus the AI-generated elements. We saw that comparison for Total Panel, but just for the BDV (binary-dependent variable) R54X. Now we see the same analysis, but for all the binary transformed variable.

The coefficients for the human-generated elements are displayed in Table 7.13. The coefficients for the AI-generated elements are displayed in Table 7.14. Tables

Table 7.13 Coefficients for binary-dependent variables and RT, for human-generated elements

		Binary dependent variables, and response time (RT)									
	Human-generated elements	**R54X**	**R52X**	**R21X**	**R41X**	**R1X**	**R2X**	**R3X**	**R4X**	**R5X**	**RT**
A1	I believe that students from different backgrounds have equal access to quality education	17	13	4	8	0	4	7	8	9	1.0
A2	I do not think students from different backgrounds have equal access to quality education	17	13	4	7	1	3	8	7	10	1.0
A3	The access to quality education for students from different backgrounds varies depending on several socioeconomic factors	18	9	2	11	1	1	7	10	9	1.0
A4	While there are still disparities, the situation is improving, and we are moving towards more equal access to quality education for students from different backgrounds	18	13	4	8	-1	4	8	8	9	1.0
B1	The effectiveness of teaching methods, curriculum design, and assessment practices are indispensable in shaping student success	18	17	5	7	3	2	8	4	14	1.2

C3	The impact of technology in education can vary depending on the specific context and how it is implemented	18	16	8	10	3	5	6	7	10	1.2
C4	Overreliance on technology may lead to concerns about screen time, privacy, and equity issues	18	15	7	10	1	5	4	8	10	1.0
D1	Developing fair and reliable assessment methods is essential for students' success	18	11	2	9	2	0	8	7	11	0.7
D2	Overemphasis on standardized testing may lead to teaching to the test and narrowing the curriculum	15	10	4	8	3	1	8	5	9	0.8
D3	The importance of assessment methods can vary depending on the educational context and goals	17	12	2	7	0	2	7	7	10	0.7
D4	Overvaluing assessment methods can create undue stress for students and teachers	18	9	1	10	3	-2	8	7	11	1.0

Tab. 7.13 (continued)

B2	Factors such as student motivation, home environment, and access to resources also play crucial roles in students' success	19	17	5	7	1	4	6	6	12	1.2
B3	The impact of teaching methods, curriculum design, and assessment practices on student success can vary widely depending on the context and the specific needs of students	15	16	6	5	2	4	10	3	12	1.2
B4	Factors such as parental involvement, socioeconomic status, and access to educational resources are determinant in students' success	21	14	3	9	1	2	7	8	13	1.1
C1	The integration of technology in education is crucial for students' success	18	15	6	10	1	5	6	9	10	1.1
C2	Technology can be beneficial in the education process, but its importance should not be overstated	17	16	6	7	1	6	6	7	11	1.0

Tab. 7.13 (continued)

Table 7.14 Coefficients for binary-dependent variables and RT, for AI-generated elements

		Binary Dependent Variables and Response Time (RT)									
	AI-generated Elements	**R54X**	**R52X**	**R21X**	**R41X**	**R1X**	**R2X**	**R3X**	**R4X**	**R5X**	**RT**
A1	Students with Down's syndrome, and those with mental retardation, autistic and slow learners, are integral parts of HEI auditors	**23**	14	2	11	0	1	4	11	12	1.0
A2	Students of wealthy or not wealthy families and those of vulnerable groups have the same access to HEI	**22**	17	4	8	2	2	3	6	15	1.0
A3	The number of international students in HEIs is continuously increasing	**24**	14	1	11	1	1	3	10	14	0.9
A4	HEI motivates dialogue, negotiation, cultural and religious tolerance in its study programs and research projects	**25**	17	4	12	2	2	0	10	15	1.0
B1	Professors implement teaching methods that capture students' attention and develop their interests	**21**	14	2	9	1	1	8	8	14	1.0
B2	HEI frequently trains professors concerning innovative teaching methods	19	18	4	4	1	2	7	3	16	1.1
B3	HEI implements student-centered methods, project-based teaching,	18	18	4	4	1	3	9	3	14	1.2

	and collaborative learning in all programs										
B4	Students interact with each other in the classroom for problem-solving issues	20	19	3	4	1	2	8	3	17	0.9
C1	HEI motivates the students toward critical thinking through implementation of various teaching and learning technologies	16	17	6	5	2	4	8	3	13	0.7
C2	The implementation of innovative teaching and learning motivates students to creativity	20	17	5	8	2	3	5	6	14	0.6
C3	The implementation of teaching and learning technologies improves learning outcomes	18	11	6	13	3	3	8	10	7	0.8
C4	The implementation of teaching technologies helps professors to understand students' needs better	16	15	6	8	3	4	7	5	11	0.7
D1	HEI assesses students based on practical works, research projects, and technical exams	18	12	5	10	4	1	5	7	11	1.0
D2	HEI uses traditional and technological student assessment methods	22	12	3	13	4	-1	2	9	13	0.8

Tab. 7.14 (continued)

D3	HEI encourages students to evaluate professors' teaching methods through semestral questionnaires	21	12	2	11	2	0	3	9	12	0.8
D4	HEI evaluates professors' didactic and scientific performance every year	19	14	2	7	3	-1	6	4	15	1.1

Tab. 7.14 (continued)

display cells that have been shaded to highlight the underlying patterns. The following are the operationally specified shading criteria to call attention to key elements. Keep in mind that Tables 7.13 and 7.14 deal with the elements, themselves, based on the deconstruction of the vignettes by OLS regression.

R54x—Shaded when the coefficient is 21 or higher.
BDV (except for R54x)—Shaded when the coefficient is 15 or higher.
RT—Shaded when the response time is 1.2 or higher.

7.7 How Well Did We Do? The IDT (Index of Divergent Thought)

It is clear just by looking at Tables 7.13 and 7.14 that the elements created by human beings generated higher coefficients than those created by AI. Our first conclusion, therefore, is that the human elements generate stronger responses. However, there is a deeper pattern to the data, which may cast a shadow on the clearly stronger performance of the elements across mind-sets. This is the conclusion reached after computing the IDT, the index of divergent.

The IDT was used to understand how well the Mind Genomics effort was used to learn about people's thoughts (Thakral et al. 2021; Madore et al. 2016). The original notion was that it was better to have high coefficients rather than low coefficients, for the obvious reason that the high coefficients represent a "strong signal." We are accustomed to search for strong effects in research, for applauding data which has "high statistical significance" in the language of the statistician working with the researcher.

Having all high coefficients for key subgroups may be something to applaud, but not in the case of mind-segments. The ideal for mind-set segmentation is high coefficients for elements in one mind-set and low coefficients for the same elements in the other mind-set(s).

Table 7.15 Calculation of the IDT (index of divergent thought)

Human-generated elements	**Total**	**MS 1 of 3**	**MS 2 of 3**	**MS 3 of 3**	**MS 1 of 2**	**MS 2 of 2**
Base	100	39	31	30	55	45
Sum Total of Squares of All Positive Coefficients	5000	5994	6998	5047	5945	4683
Average of All Positive Coefficients (sum coefficients/base)	50	154	216	168.2	108	104
Number of respondents in the subgroup)	100	39	31	30	55	45
Weight = Proportion of the respondents in the subgroup	0.33	0.13	0.10	0.10	0.18	0.15
Weighted Total (Weight x Sum Total of Squares of All Positive Coefficients)	1650	779	670	505	1070	702
Final Score = sum of weighted total and its squared root, the IDT of 73, suggesting strong differentiation of mind-sets	5376	73				
AI-generated elements.	**Total**	**MS 1 of 3**	**MS 2 of 3**	**MS 3 of 3**	**MS 1 of 2**	**MS 2 of 2**
Base (number of respondents in this group)	100	31	40	29	47	53
Sum Total of Squares of All Positive Coefficients	6586	7142	6661	7634	6357	7086
Average of All Positive Coefficients (sum coefficients/base)	66	230	167	263	135	134
Number of respondents in the column (in the subgroup)	100	31	40	29	47	53
Weight = Proportion of the respondents in the subgroup	0.33	0.10	0.13	0.10	0.16	0.18
Weighted Total (Weight x Sum Total of Squares of All Positive Coefficients)	2173	714	866	763	1017	1275
Final Score = sum of weighted total and its square root, the IDT of 83 suggesting strong performance of elements but poorer differentiation of mind-sets	6810	83				

The IDT was suggested as a computational formula that showed the performance of the ideas. The computational formula is shown at the top of Table 7.15 for the human-generated elements and at the bottom of Table 7.15 for the AI-generated elements. The IDT measures the elements' performance (Giancola et al. 2023). Low IDT numbers correspond to elements, which perform poorly across all subgroups and are not relevant. In contrast, high IDT values correspond to elements that performed well across all mind-sets, at first a seemingly positive results, but not after further thinking about what the IDT tells us. High IDT values suggest that all mind-sets respond strongly to the elements, something which is not desirable when we divide people into mind-sets. With mind-sets, we want people to react strongly only to different, nonoverlapping sets of elements.

Simulations suggest that IDT values around 70 correspond to strong performing elements in some mind-sets and not in others, a finding which is more meaningful than observing strong performing elements in all mind-sets. Based upon the newly developed rule of thumb that IDT values of 70 or so are reflections of stronger segmentations and more dissimilar clustering, we are left with the interesting results that the IDT value for human-generated elements is 73. In contrast, the IDT value for AI-suggested elements is 83. It is the human being that somewhat out-thinks the AI, at least so far!

7.8 Conclusions

This chapter compares the human response to ideas about education, with sources of ideas from humans versus AI, but with the same silos or questions. Only the answer differs. The objective was to determine whether the source of the ideas makes a difference. The data suggests that both sources of ideas generate strong elements, but only when the human respondents are divided into mind-sets by standard clustering methods used in Mind Genomics. The most important finding is that the AI effort can end up being "too good," generating in its wake mind-sets, but with the mind-sets generated from AI elements being simply different groups both liking most of the elements. It is when the elements emerge from human minds that we find much more conventional mind-set segmentation, viz., that some people are really interested in the ideas, whereas others are indifferent to the same ideas.

A second result from this study is the introduction of "summarization by AI," after the study is run. AI can be programmed to focus on strong-performing elements, whether for total panel or for key subgroups, such a mind-set. With a battery of prompts the AI can play the role of an independent analysis, able to find nuances in the data, which would elude people. The ability to obtain summarization within 15–30 min after the study is done and completely analyzed by the Mind Genomics platform provide a new capability to understand minds and how they work in the world of everyday life. Summarizers provide a rich set of recommendations when analyzing an input set. These recommendations are valuable as they present points of view that are not always visible to the human eye, thus essentially completing the human arguments.

References

Amaral, A. (2022). Equity in Higher Education: Evidences, Policies and Practices. Setting the Scene. In *Equity Policies in Global Higher Education* (pp. 23–46). Springer International Publishing. https://doi.org/10.1007/978-3-030-69691-7_2

Buchholtz, N., Stuart, A., & Frønes, T. S. (2021). Equity, equality and diversity—Putting educational justice in the Nordic model to a test. In T. S. Frønes, A. Pettersen, J. Radišić, & N. Buchholtz (Eds.), Equity, equality and diversity in the Nordic model of education (pp. 13–41). Springer. https://doi.org/10.1007/978-3-030-61648-9_2

Carvalho, A. E., Cosme, A., & Veiga, A. (2023). Inclusive education systems: The struggle for equity and the promotion of autonomy in Portugal. *Education Sciences, 13*(9), 875. https://doi.org/10.3390/educsci13090875

Chugh, R., Turnbull, D., Cowling, M. A., Vanderburg, R., & Vanderburg, M. A. (2023). Implementing educational technology in Higher Education Institutions: A review of technologies, stakeholder perceptions, frameworks and metrics. *Education and Information Technologies*, *28*(12), 16403–16429. https://doi.org/10.1007/s10639-023-11846-x

Dawson, P., & McHenry, C. R. (2022). Approaches to Assessment: A Perspective from Education. In *Clinical Education for the Health Professions* (pp. 1–13). Springer Singapore. https://doi.org/10.1007/978-981-13-6106-7_74-1

Giancola, M., D'Amico, S., & Palmiero, M. (2023). Working memory and divergent thinking: The moderating role of field-dependent-independent cognitive style in adolescence. *Behavioral Sciences, 13*(5), Article 397. https://doi.org/10.3390/bs13050397

Gofman, A. (2012). Origins of RDE and the role of experimentation in consumer-driven innovation. In A. Gofman & H. R. Moskowitz (Eds.), Rule developing experimentation: A systematic approach to understand & engineer the consumer mind (pp. 3–24). *Bentham Science Publishers*. https://doi.org/10.2174/978160805284411201010003

Ilollari, O., Papajorgji, P., & Civici, A. (2020). Tourism in Albania and its challenges during and after Covid-19. Proceedings of the 3rd UNICART Interdisciplinary International Conference.

Jurado de los Santos P., Moreno-Guerrero A.-J., Marín-Marín J.-A., & Soler Costa R. (2020). The Term Equity in Education: A Literature Review with Scientific Mapping in Web of Science. *International Journal of Environmental Research and Public Health*, *17*(10), 3526. https://doi.org/10.3390/ijerph17103526

Madore, K. P., Jing, H. G., & Schacter, D. L. (2016). Divergent creative thinking in young and older adults: Extending the effects of an episodic specificity induction. *Memory & Cognition, 44*(6), 974–988. https://doi.org/10.3758/s13421-016-0605-z

Milutinovic, V., & Salom, J. (2016). Mind genomics: A guide to data-driven marketing strategy. Springer. https://doi.org/10.1007/978-3-319-39733-7

Moskowitz, H. R., Gofman, A., Beckley, J., & Ashman, H. (2006). Founding a new science: Mind genomics. *Journal of Sensory Studies, 21*(3), 266–307. https://doi.org/10.1111/j.1745-459X.2006.00066.x

Mucherino, A., Papajorgji, P., & Pardalos, P. (2009). Clustering by k-means. In *Data Mining in Agriculture* (pp. 47–82).

Noben, I., Deinum, J. F., & Hofman, W. H. A. (2022). Quality of teaching in higher education: reviewing teaching behaviour through classroom observations. *International Journal for Academic Development*, *27*(1), 31–44. https://doi.org/10.1080/1360144X.2020.1830776

OpenAI. (2024). *ChatGPT [Large language model]*. https://chat.openai.com

Prodan, A., Maxim, E., Manolescu, I., Arustei, C. C., & Guta, A. L. (2015). Access to Higher Education: Influences and Possible Implications. *Procedia Economics and Finance*, *20*, 535–543. https://doi.org/10.1016/S2212-5671(15)00106-9

Rawlsyk, P. E. (2018). Assessment in Higher Education and Student Learning. *Journal of Instructional Pedagogies*, *21*.

Sokoli, D., Trunk Širca, N., & Koren, A. (2021). Quality of teaching in Kosovo's higher education institutions: Viewpoints of institutional leaders and lecturers1. *Human Systems Management*, *40*(5), 685–700. https://doi.org/10.3233/HSM-201155

Thakral, P. P., Yang, A. C., Addis, D. R., & Schacter, D. L. (2021). Divergent thinking and constructing future events: Dissociating old from new ideas. *Memory, 29*(6), 729–743. https://doi.org/10.1080/09658211.2021.1940205

Todri, A., Papajorgji, P., Moskowitz, H., & Scalera, F. (2021). Perceptions regarding distance learning in higher education, smoothing the transition. *Contemporary Educational Technology, 13*(1), Article ep287. https://doi.org/10.30935/cedtech/9274

Vasconcelos, C., & Calheiros, C. S. C. (2022). *Enhancing Environmental Education Through Nature-Based Solutions*. Springer International Publishing.

Wanti, M., Wesselink, R., Biemans, H., & den Brok, P. (2022). Determining factors of access and equity in higher education: A systematic review. *Equity in Education & Society, 1*(2), 279–296. https://doi.org/10.1177/27526461221092429

Chapter 8
Experiment in Family Stress Issues

8.1 Introduction to Family Stress Model (FSM)

Family stress is all too common. The issues are many, resulting from the interplay of the many factors involved in daily life. Furthermore, stress is something to be studied in real life, not something that can be induced by the researcher. Inducing stress would be unethical to say the least. Furthermore, with the tools of Mind Genomics explicated below, it will be straightforward to measure some aspects of family stress, such as the degree to which readers are engaged when reading about the different aspects of family stress.

Stress can include family financial difficulties, work-related stress, health issues, relationship problems, parenting challenges, and the impact of external factors like societal changes or crises such as pandemics and economic downturns. Which, however, are most important to a person? Can we develop a model of the mind, telling us which specific aspects of stress focus a person, and engage the person's attention?

Investigations of family stress is not new, indeed the opposite. Literature in the field going back 75 years focused on the nature of different variables that interact to product stress. This pathbreaking effort is known as the Family Stress Model (Hill 1949). At the core is the assumption that families are not victims of their circumstances. Rather, families are dynamic organisms. They take advantage of resources available to them, and with the proper attitude, they can adapt in a positive fashion, becoming resilient in the face of their challenges. They can perceive meaning in their experience; instead, as they take advantage of resources and perceive meaning in their experiences, families can positively adapt and be resilient in the face of their challenges.

The model will become important as the source of the test stimuli for the Mind Genomics study discussed in this chapter.

P. Papajorgji, H. Moskowitz, *The Mind of Everyday*,
https://doi.org/10.1007/978-3-031-78078-3_8

1. **Stressors,** events or situations which disrupt family equilibrium, such as illness, job loss, or divorce.
2. **Family resources,** internal strengths and external supports available to the family such as communication skills, family financial stability, and social networks.
3. **Perception of the stressor**, meaning how the family interprets and appraises it, impacting their emotional and cognitive responses.
4. **Coping mechanism:** To successfully overcome stress, it is crucial to define a set of strategies to manage stress, such as the ability to seek support, family problem-solving skills, and emotional regulation.
5. **Adaptation.** The last element of FSM is the *adaptation* phase, representing the resulting outcome of the family's interaction with the stressor, ranging from favorable growth to maladaptive responses (McCubbin and Patterson 1983; Heyl 2022).

The FSM was a pioneering work and offered a foundational framework for understanding family stress reactions. It had some limitations; it oversimplified the complexity of family dynamics and lacked detail on specific processes (McCubbin and Patterson 1983). The developed model, later called FSP, has been used to investigate different aspects of family stress (Hill et al. 2017; Zietz et al. 2022; Neppl et al. 2016), as well as adolescent problematic behaviors (Chen et al. 2024).

The topic of family stress generated an increasing amount of research, understandably so because of the importance of the topic for everyday life. Later, Hill tried to converge parallel developments in this field and define the generic features of families under stress. The expanded model is known as the ABCX model (Hill 1949). The ABCX model subsequently evolved to the Double ABCX Model of Adaptation, focusing on the many factors affecting a family's adjustment to a severe ongoing stressor (McCubbin and Patterson 1983). The ABCX model includes the family's adjustment to the cumulative effects of the stressor, after the initial or original crisis. The original ABCX and the expanded Double ABCX models identify variables that account for observed differences among families' adaptations to stressful situations (Cherry 1989).

The topic of family stress remains important for research, with studies of all types. Studies have been published dealing with the connections between marital distress, harsh parenting, and subsequent child externalizing behaviors in line with predictions from the Family Stress Model (FSM) (Neppl et al. 2016). The Family Stress Model of Economic Hardship (FSM) asserts that economic situations create differences in psychosocial outcomes for parents and developmental outcomes for their adolescent children (Zietz et al. 2022). Masarik and Conger (2017) provide an illustrative review of the FSM framework to understand how family stress impacts children across development in physical, social-emotional, and cognitive domains.

It is clear from the array of studies and the interaction of variables that we have in the topic of family stress a socially meaningful topic area, one that might benefit from the Mind Genomics approach as a simplifying procedure. As family stress issues play a significant role in family well-being, we decided to study what impact the use of AI will have in conducting scientific research in this field. The specific idea underlying the research of this chapter is the study of how people respond to different facets of family stress in the Mind Genomics paradigm, and then the response to parallel types of "test message or elements" when these elements are created by human beings, with an emotional sensibility versus when these elements are created by AI, which is not (as yet) known for its emotionality.

8.2 The Mind Genomics Experiments

The Mind Genomics process is simple and templated. The steps are set up so that the studies can be run even by people with little experience. The Mind Genomics platform is BimiLeap.com. The system is set up with prompts, and with both AI as a coach to help the researcher create elements, as well as with an automated analysis system once the researcher has done the study with human respondents.

Step 1: Create four questions to be answered, or in the language of Mind Genomics four silos or categories. These will give structure to the test stimuli, but will never themselves be seen (Kover et al. 2022). It is from the literature that these four questions or silos emerge. The source of the questions/silo come from abstracting the large amount of information currently published (Cherry 1989; McCubbin and Patterson 1983; Hill 1949; Hill et al. 2017). Table 8.1. presents these four silos or questions. They will be used in the two Mind Genomics studies, the first for human-generated messages (elements), and then for AI-generated messages.

Step 2: The questions or silos for both studies are the same. What is different is the way the elements for each silo are determined. In the first study, the elements for each silo were selected by human beings, the researchers. In the second study the elements were determined by AI. The AI (Chat GPT3.5) is embedded in the Mind Genomics platform and called the Idea Coach. The researcher simply typed in the question to the Idea Coach and requested answers. The AI returned with sets of 15 answers for each question and could iterate for any question several times. After one or two iterations for each question the researcher generated the requisite four answers. Table 8.2 shows the silos and the elements for the human-based study. Table 8.3 shows the silos and elements for the AI-based

Table 8.1 The four silos (questions) for the two experiments

Question A: Economic-based stress: Do you think economic issues are the main reason for the stress in the family?
Question B: Health Issues: Do you think health issues are the main reason for the stress in the family?
Question C: Family Relationships: Do you think family relationships are the main reason for the stress in the family?
Question D: Family members dependencies: Do you think addiction to bad habits, drugs, alcohol are the main reason for the stress in the family?

Table 8.2 Silos and the corresponding elements for the human-based study

Human-created elements	
Question A: economic-based stress: do you think economic issues are the main reason for the stress in the family?	
A1	I think economic-related issues are the main reason for the stress in the family
A2	I do not think the economic-related issues are the main reason for the stress in the family
A3	There are many reasons other than economic issues to have stress in the family
A4	Our family does not have economic issues
Question B: health issues: do you think health issues are the main reason for the stress in the family?	
B1	I think health-related issues are the main reason for the stress in the family
B2	I don't think health-related issues are the main reason for the stress in the family
B3	There are many reasons other than health issues to have stress in the family
B4	There are many reasons other than health issues to have stress in the family
Question C: family relationships: do you think family relationships are the main reason for the stress in the family?	
C1	I think family relationship issues are the main reason for the stress in the family
C2	I do not think family relationship issues are the main reason for the stress in the family
C3	There are many reasons other than family relationship issues to have stress in the family
C4	Our family does not have family relationships issues
Question D: family members dependencies: do you think addiction to bad habits, drugs, alcohol are the main reason for the stress in the family?	
D1	I think addiction to bad habits, drugs, alcohol are the main reason for the stress in the family
D2	I do not think addiction to bad habits, drugs, alcohol are the main reason for the stress in the family
D3	There are many reasons other than addiction issues to have stress in the family
D4	Our family does not have addiction-related issues

study. The elements were slightly polished to be "stand-alone" phrases, a requirement for the Mind Genomics study.

Step 3: Create the test stimuli. The test elements (answers) are put into combinations, comprising two, three, or four elements. Each combination is called a vignette. A vignette has at most one element or one answer from each of the four questions. It will be these vignettes that a person evaluates in the study. Each person will end up evaluating 24 of these vignettes. The vignettes are constructed

Table 8.3 Silos and the corresponding elements for the AI-based study

AI-Generated elements	
Question A: Economic-based stress: Do you think economic issues are the main reason for the stress in the family?	
A1	Economic Issues are the Primary Source of Family Stress
A2	Economic Issues are Just One of Many Factors of Family Stress
A3	Economic Issues are Stress depending on the Context or Stage of Life
A4	Economic Stress is a Symptom of Deeper Issues in Family
Question B: Do you think health issues are the main reason for stress in the family?	
B1	Health Issues are the Primary Source of Family Stress
B2	Health Issues are Important but Not the Only Family Stressors
B3	Health Issues can be Stress Depending on Individual and Family Circumstances
B4	Health Issues are Both Cause and Effect of Family Stress
Question C: Do you think family relationships are the main reason for stress in the family?	
C1	Family Relationships are the Primary Source of Stress
C2	Family Relationships are Just One Aspect of Stress
C3	Family Relationships can be Stress Depending on the Family's Situation and Dynamics
C4	Family Relationships are Both Cause and Effect of Family Stress
Question D: Do you think addiction to bad habits, drugs, or alcohol are the main reason for stress in the family?	
D1	Addictions are the Primary Source of Family Stress
D2	Addictions are One of Many Family Stressors
D3	Addictions cause Family Stress Depending on the Severity and Impact of the Addiction
D4	Addictions are Both Cause and Consequence of Family Stress

by what is called an experimental design. The experimental design ensures that for each set of 24 combinations or vignettes evaluated by a respondent, the 16 elements will appear in what is called a statistically independent fashion. The important thing to know in a Mind Genomics study is that each respondent, each survey participant, evaluate 24 ***different v***ignettes. This property of the vignettes is ensured by a process called permuted experimental design (Gofman and Moskowitz 2010). The idea of the permuted design is very similar to the idea underlying an MRI. Rather than having a hundred people evaluate the same set of 24 combinations, we have a hundred people evaluating 24 sets or 2400 different combinations. In so doing, we cover a wide space of combinations, making it more likely that the research will discover meaningful patterns.

An important thing to note in these studies is that the experimental design, which underlies the combinations, ensures that the data can be analyzed with any combination of people, even the data from one person at a time. That experimental design will allow us to explore individual patterns and allow us to cluster people into mindsets. That analysis will be discussed later. The final thing to keep in mind is that the Mind Genomics approach differs from the typical research approach of hypothesis testing and confirmation. The results end up with the discovery of these patterns, whether or not we deeply understand the topic.

Step 4: Create the rating scale. The typical rating scale used by a researcher in these types of studies is called the Likert scale. The scale usually comprises one dimension, such as a 1 to 5 scale, or a 1 to 7 scale, or a 1 to 9 scale. The scale has one dimension, such as *would not buy to would buy*, or *does not affect me versus affects me,* and so forth.

In recent years, the Mind Genomics effort has focused on a two-sided scale, allowing the researcher to obtain more information from a single judgment. One side of the scale might be emotional, e.g., does what I read interest me, yes or no? The other side of the scale might be *do I believe what I read, or will it happen, yes or no?*

It is this type of two-sided scale that we used here. Table 8.3 shows the five scale points. Notice that this scale goes from 1 to 5. From the point of view of the respondent the left-hand side of the scale is whether I'm interested or not interested in what I'm reading about the family stress. In turn, the right-hand side of the scale tells me whether I believe it will happen or not. We have thus a very compact way of having the respondent rate the vignette two dimensions, the first dimension emotional, the second dimension intellectual. Table 8.4 shows the scale as presented to the respondent.

Step 5 – Run the study on the Internet. Running a Mind Genomics study is straightforward as we have seen in the previous chapters. Once the researcher puts the elements and the rating scale into the template, the program takes over. The respondent requests a certain sample, generally balanced for age and gender. The program provides a link to the study. The researcher sends out the link to prospective participants, or has a service send out the link. Special care was paid to ensure a good distribution of parameters such as age, education level, and gender by carefully selecting the participants in the study.

Participants who agreed to do the study pressed the link and were led to the study itself. The study begins with an orientation shown below.

Table 8.4 The five-point, two-sided Likert scale used to evaluate each vignette

1 = Not interested ... AND ....won't happen
2 = Not interested ... BUT .. it could happen
3 = Can't honestly say
4 = Looks interesting .. BUT...won't happen
5 = Looks interesting .. AND .. it could happen

> ***Stress can have severe consequences for the family. Several reasons, such as economics, health, addictions, and family relationships can cause stress. We would like to have your opinion to understand and address these issues.***

Following the orientation, the respondent sees 24 different vignettes, namely, the 24 vignettes from a single one of the permuted experimental design. The vignettes themselves are very simple. They comprise a short introduction of a few words, the vignette, and the scale. The "spare" look makes the respondent's task easier. There is less to read. The respondent simply looks at the vignette and assigns a rating. The Mind Genomics system is impossible to "game." Each respondent sees what seems to be a random combination of elements on a vignette and has a hard time knowing the correct answer. Most respondents stop trying to "give the correct answer," and end up guessing, which is exactly what is desired. Typically, the respondent assigns a rating based upon what she or he thinks is the right answer. This actually happens very quickly, with each vignette rated in less than 2–4 s. At the end of the evaluation, the respondent finishes and the data are sent back for analysis. Exit interviews with many respondents come up with the same statement, namely, the respondent thought that it was very confusing. They stopped trying to guess what the structure was or what the right answer should be and simply answered appropriately almost immediately, which is precisely what was desired.

<u>Step 6 – Transform the ratings to allow analysis using OLS (ordinary leas-squares) regression</u>. Although the study involved the evaluation of vignettes, the real analysis comes from the study of the performance of the individual elements, one set for those developed by human beings and the other developed by artificial intelligence.

The analysis uses OLS (ordinary least squares regression) to relate the presence/absence of the 16 elements to the rating. The five-point scale that we created cannot be easily used for regression. There are two reasons for this. The first reason is that the scale has two dimensions (interesting, likely to happen), whereas the regression can only have one dimension. The second reason is that those who use the results typically do not understand what 1, 2, 3, 4, and 5 is. When presented with the results of an average of 3.4 from a 5-point scale one will undoubtedly ask about the meaning of this average. When one hears questions of this nature, it's time then to change the analysis to something that the manager can understand.

The analysis transformation prior to the analysis appears in Table 8.5. With the binary scale we can create a variety of scales as shown in Table 8.5. We will primarily focus on the scale, which is the ratings of 5 and 4 transformed to 100 and the rating of 1, 2, and 3 transformed to 0. This is the interest scale. We are focusing on the relation between the presence of each of the 16 elements in the experiment and the selection of "Interests Me." Will we find differences among the elements in a study in terms of "Interests Me," and even more important, will we find differences when we switch from the human being providing the elements to AI providing the elements?

Table 8.5 Transformation of two-side Likert Scale values to 0 or 100 (binary transformed value)

	The original points on the rating scale				
	1	2	3	4	5
Left side of scale: interests me	N	N	DK	Y	Y
Right side of scale: happen	N	Y	DK	N	Y
Newly transformed binary variables					
R1x	100	0	0	0	0
R2x	0	100	0	0	0
R3x	0	0	100	0	0
R4x	0	0	0	100	0
R5x	0	0	0	0	100
R54x	0	0	0	100	100
R52x	0	100	0	0	100
R41x	100	0	0	100	0
R21x	100	100	0	0	0

One other thing should be noted, the protection against the problem of having all of the ratings of a person transformed either to 0 or to 100. For example, if a person always chose 5 or 4 to rate the 24 vignettes, that person would generate 24 values of "100" for the newly transformed binary variable R54x (see Table 8.5). To protect against this possibility, a situation which would stop the OLS regression, we add a vanishingly small random number ($<10^{-5}$) to every transformed binary variable. The random number has no material effect on the analysis but ensures that there is the requisite variation in all of the newly transformed binary variables.

Step 7 – How do the respondents evaluate the vignettes in the two studies? The percentages of rating points allotted to the vignettes are examined in the first "morphological" analysis. The main point of interest lies in the distinction between the rating distributions of elements selected by AI and those selected by humans. Once more, the only information available regarding the individual pieces is what their sources say. The frequency of ratings of "5" (interest me; likely to happen) is higher for the elements generated by human beings in comparison to elements generated by AI. This discovery suggests a difference in the deep nature of elements (really information) generated by humans versus by AI. Table 8.6 shows this comparison.

Step 8 – How do the elements "drive" the binary transformed rating? Mind Genomics focuses on the individual elements, specifically trying to quantify the contribution of each element to the rating. Recall that the experimental design mixes and matches the individual elements when it creates the vignettes. We need a method by which to deconstruct the rating into the part worth contribution of each of the 16 elements. The method is called OLS (ordinary least squares) regression.

The equation is written as follows: We express the equation as: $R54x = k_1A1 + k_2A2 \ldots k_{16}D4$. Each of our 16 elements ends up with a "weight," the coefficient. The transformed rating that we created is a weighted combination of

Table 8.6 Frequency (percent) of selection across the 2400 vignettes for the human-sourced elements and for the AI-sourced elements

	Percent of time select single rating point					Percent of time select interesting or not interesting		Percent of time select– likely to happen or not likely to happen		Response time
Interest me	**N**	**N**	**NA**	**Y**	**Y**	**Yes**	**No**			
Likely to happen	**N**	**Y**	**NA**	**N**	**Y**			**Yes**	**No**	
Original rating	**1**	**2**	**3**	**4**	**5**	**5 or 4**	**1 or 2**	**5 or 2**	**1 or 4**	
	R1X	**R2X**	**R3X**	**R4X**	**R5X**	**R54X**	**R12X**	**R52X**	**R41X**	**RT (Sec)**
Total	7	12	24	295	29	57	19	41	35	3.7
Source – Human	8	14	25	27	26	53	22	40	35	4.1
Source – AI	5	11	22	30	32	62	16	42	36	3.3

the 16 different elements. The method of ordinary least squares regression will estimate the contribution for us. We use the regression to estimate the weights. Note that there is no additive constant. This is called *forcing the equation through the origin*. The reason for doing so is that all the contributory power is shown by the elements. Once we do the regression analysis, we can look immediately at the coefficients, k_1 to k_{16}, and learn how each element drives the rating.

For our analysis we choose one of the newly created binary variables (see Table 8.6). That variable is R54x, the one which becomes 100 when the rating is 5 or 4 (interested), or 0 when the rating of 3,2, or 1 (not interested).

The important thing emerging from the OLS regression is the magnitude of the coefficient. When the coefficient is 21 or higher, experience suggests that this is an important characteristic, an important statement, or one which looks interesting. When the coefficient is 20 or less, we operationally define that as not being as interesting. It happens to be that around 20 or so is a very strong statistically significant effect. But we're not interested here in whether there's a statistically significant difference from zero for the coefficient, but rather simply how strong. And we need a cutoff point to tell us to look at these elements and not to look at those.

When we look at Table 8.7, the elements generated by the human beings, we find only one element, which ends up driving R54x, viz., interested. This is "There are many reasons other than health issues to have stress in the family." The coefficient is only 21. When we move to Table 8.8, the elements generated by AI, we find six of the 16 elements to perform strongly, four from economic issues, and two from health issues.

The remarkable effect here is that the AI-generated elements are far more interesting than are the human-generated elements. The interesting question is "why." One possible reason may be that the AI-generated elements are simple to read, and

Table 8.7 Coefficients for human-generated elements for family stress The dependent variable is R54x (looks interesting)

Human-Generated Elements for Family Stress Dependent Variable R54x – looks interesting	
Base (number of respondents in this group)	**100**
There are many reasons other than health issues to have stress in the family.	21
I think family relationship issues are the main reason for the stress in the family.	18
I do not think family relationship issues are the main reason for the stress in the family.	18
There are many reasons other than economic issues to have stress in the family.	17
Our family does not have family relationships issues.	17
I think economic-related issues are the main reason for the stress in the family.	16
I don't think health-related issues are the main reason for the stress in the family.	16
There are many reasons other than health issues to have stress in the family.	16
There are many reasons other than family relationship issues to have stress in the family.	16
Our family does not have addiction-related issues.	15
Our family does not have economic issues.	14
I think health-related issues are the main reason for the stress in the family.	14
I think addiction to bad habits, drugs, alcohol are the main reason for the stress in the family.	14
There are many reasons other than addiction issues to have stress in the family.	14
I do not think the economic-related issues are the main reason for the stress in the family.	11
I do not think addiction to bad habits, drugs, alcohol are the main reason for the stress in the family.	10

impactful. The silo appears, followed by a very succinct and meaningful statement, easy to understand.

8.3 AI Summarization of Strong Performing Elements

As part of the Mind Genomics platform, an AI-based summarizer using Chap GPT3.5 was used to explore different aspects of strong performing elements, scoring 21 or higher. The summarizer is automatically performed only on those

Table 8.8 Coefficients for AI-generated elements for family stress The dependent variable is R54x (looks interesting)

AI-Generated Elements for Family Stress – Dependent Variable R54x (Looks interesting)	
Base (number of respondents in this group)	**100**
Economic Stress is a Symptom of Deeper Issues in Family	25
Economic Issues are Just One of Many Factors of Family Stress	24
Economic Issues are the Primary Source of Family Stress	22
Economic Issues are Stress depending on the Context or Stage of Life	22
Health Issues can be Stress Depending on Individual and Family Circumstances	22
Health Issues are Both Cause and Effect of Family Stress	22
Health Issues are the Primary Source of Family Stress	20
Family Relationships are the Primary Source of Stress	19
Health Issues are Important but Not the Only Family Stressors	16
Family Relationships can be Stress Depending on the Family's Situation and Dynamics	16
Family Relationships are Just One Aspect of Stress	15
Family Relationships are Both Cause and Effect of Family Stress	15
Addictions cause Family Stress Depending on the Severity and Impact of the Addiction	14
Addictions are Both Cause and Consequence of Family Stress	13
Addictions are One of Many Family Stressors	12
Addictions are the Primary Source of Family Stress	11

Table 8.9 AI-summarized results for the human-generated elements

AI summarization of human-generated elements, total panel
Interested in:
There are many reasons other than health issues to have stress in the family.
Create a label.
"Other Factors Contributing to Family Stress"
Describe this segment:
This segment of the research study includes individuals who believe that there are numerous factors contributing to stress within families, aside from health-related problems. These individuals may acknowledge that stress can stem from various sources such as financial difficulties, relationship conflicts, career pressures, or other external factors. They recognize that health issues are not the sole cause of stress within families and may be more inclined to explore and consider these alternative reasons. This segment's perspective suggests a broader understanding of the complexities and multifaceted nature of family stress.

elements. Table 8.9 summarizes the results for the Human-Generated elements. Table 8.10 summarizes the results for the AI-generated elements.

When the AI summarizer works on the strong performing elements generated by humans, the output is remarkably shorter than the summarization generated from considering the elements generated by AI. One possible reason is that the elements generated by human beings incorporate many different emotions, feelings, shades

Table 8.10 AI-summarized results for the AI-generated elements

AI summarization of AI-generated elements, total panel
Interested in:
Economic Issues are the Primary Source of Family Stress; Economic Issues are Just One of Many Factors of Family Stress; Economic Issues are Stress depending on the Context or Stage of Life; Economic Stress is a Symptom of Deeper Issues in Family; Health Issues can be Stress Depending on Individual and Family Circumstances; Health Issues are Both Cause and Effect of Family Stress;
Create a label
Family Stress Factors: Economic and Health
Describe this segment:
This segment consists of individuals who believe that economic issues are the main source of stress for families. They perceive that economic issues play a significant role in causing stress within families. However, they also recognize that economic issues are just one of several factors that contribute to family stress. They understand that the impact of economic stress may vary depending on the specific context or stage of life a family is in. Moreover, these individuals believe that economic stress is not just a standalone problem but can be a symptom of deeper underlying issues within the family.
Furthermore, this segment acknowledges that health issues can also be a source of stress for families. They understand that the level of stress caused by health issues may vary depending on individual and family circumstances. Additionally, they recognize that health issues can both be a cause and an effect of family stress, suggesting a complex relationship between health and family well-being.
Overall, this segment has a holistic understanding of the various factors that contribute to family stress, including economic and health issues. They recognize that these stressors are not isolated but interconnected, and their perception of stress is influenced by a range of individual, family, and contextual factors.

of subtly. In contrast, the elements generated by AI are direct, to the point, stating the facts, and are easier to read. There is not a subtly of meaning. Consequently, when human respondents in the study are presented with the vignettes, the elements from the AI stand out, almost as a simple list. There is no circuitous language to stand in the way of understanding. In other words, the effort by people to provide elements is diluted by the effort to phrase the elements in descriptive language. Rather than a simple and direct statement, the person creating elements is adding a layer of cognitive complexity through the nuanced presentation of the information.

8.4 Beyond the Single Dependent Variable to the Patterns for All Binary Transformed Variables and Response Time

The next analysis looks at the spectrum of coefficients across all the binary-dependent variables shown in Table 8.5. The same analysis and modeling were used. Table 8.11 shows the average values for the newly created binary-dependent variables (BDV). These averages are computed after coding the BDV 0 or 100, and

Table 8.11 The average binary-dependent variables assigned to the vignettes in both studies

Interesting	**N**	**N**	**DK**	**Y**	**Y**	**Y**	**N**			
Likely to happen	**N**	**Y**	**DK**	**N**	**Y**			**Y**	**Y**	**RT SECONDS**
Family stress	**R1X**	**R2X**	**R3X**	**R4X**	**R5X**	**R54X**	**R12X**	**R52X**	**R41X**	
Total	7	13	23	29	297	57	19	41	35	3.7
Source: Human	8	14	25	27	26	53	22	40	35	4.1
Source: AI	5	11	22	30	32	62	16	42	36	3.3

so are akin to percentages. We get a sense from the averages that the human-generated elements perform more strongly. The two most telling averages are R54x, which captures "looks interesting" and R52x, which captures "could happen." For both BDVs, there are clearly stronger performances by elements created by human beings versus elements created by AI, both judged by real people.

Table 8.11 shows the average BDV (binary-dependent variables) assigned to the vignettes in the two studies. The AI-generated elements are more interesting (higher average for R54x), are faster to process (RT values), but are no more likely to happen than are the human-generated elements. The key here is the stronger performance on interesting, and the faster comprehension of the AI-generated elements.

A deeper understanding of the AI-generated versus human-generated elements emerges after we use OLS regression to relate the newly created BDVs and response time (RT) to the presence/absence of elements. We did the analysis above. Table 8.12 shows the coefficients for the human-generated elements. Table 8.13 shows the coefficients for the AI-generated elements. The table uses shaded cells to highlight the underlying patterns. The operationally defined criteria for shading are:

R54x – Shaded when the coefficient is 21 or higher.
Remaining BDV—Shaded when the coefficient is 15 or higher.
RT—Shaded when the response time is 1.2 or higher.

The most important observations emerging from Tables 8.12 and 8.13 are that the AI-generated elements show higher coefficients than do the human-generated elements and take less time to read. AI generates easier to understand text, perhaps because the style of AI is a direct statemen, rather than a softened presentation, which requires more intellectual processing.

8.5 Moving from the Total Panel to Mind-sets

A hallmark feature of Mind Genomics is the focus on mind-sets developed from the granularity of the data and on specific topics rather than general topics. Mind Genomics works from the bottom up, looking at the pattern of responses to a specific topic, such as those studied here. From this small "slice" of reality, Mind Genomics tries to identify interpretable and presumably meaningful groups of

Table 8.12 Coefficients from the OLS regressions for human-generated elements

	Family Stress Human-Generated Elements	R54X	R52X	R21x	R41x	R1X	R2X	R3X	R4X	R5X	RT
A1	I think economic-related issues are the main reason for the stress in the family.	16	11	4	9	0	4	9	9	7	1.1
A2	I do not think the economic-related issues are the main reason for the stress in the family.	11	9	6	9	1	5	10	8	4	1.1
A3	There are many reasons other than economic issues to have stress in the family.	16	13	5	8	0	5	5	8	8	1.0
A4	Our family does not have economic issues.	13	11	7	10	3	4	7	7	6	1.1
B1	I think health-related issues are the main reason for the stress in the family.	14	10	8	11	4	4	10	8	6	1.4
B2	I don't think health-related issues are the main reason for the stress in the family.	16	13	7	10	1	6	7	9	7	1.4
B3	There are many reasons other than health issues to have stress in the family.	15	14	6	7	1	5	9	7	9	1.4
B4	There are many reasons other than health issues to have	21	13	1	9	0	2	9	10	11	1.3

Table 8.13 Coefficients from the OLS regressions for AI-generated elements

	Family Stress AI-generated elements	**R54X**	**R52X**	**R21x**	**R41x**	**R1X**	**R2X**	**R3X**	**R4X**	**R5X**	**RT**
A1	Economic Issues are the Primary Source of Family Stress	22	9	5	18	4	1	2	14	8	0.8
A2	Economic Issues are Just One of Many Factors of Family Stress	24	13	4	15	1	3	-1	14	10	0.9
A3	Economic Issues are Stress depending on the Context or Stage of Life	22	13	4	13	1	3	2	12	10	0.9
A4	Economic Stress is a Symptom of Deeper Issues in Family	25	10	3	18	1	2	2	17	8	0.9
B1	Health Issues are the Primary Source of Family Stress	20	11	4	14	1	3	7	13	8	1.0
B2	Health Issues are Important but Not the Only Family Stressors	16	12	4	8	1	4	10	7	9	1.0
B3	Health Issues can be Stress Depending on Individual and Family Circumstances	22	13	2	11	1	1	7	10	12	1.1
B4	Health Issues are Both Cause and Effect of Family Stress	22	14	4	13	2	2	5	11	12	1.0
C1	Family Relationships are the Primary Source of Stress	19	11	7	14	3	3	5	11	8	1.0
C2	Family Relationships are Just One Aspect of Stress	15	9	6	12	4	3	9	8		1.1
C3	Family Relationships can be Stress Depending on the Family's Situation and Dynamics	16	16	5	6	2	4	10	4	12	1.1
C4	Family Relationships	15	14	8	9	4	4	7	5	10	1.0

people who respond to the same elements in different ways. These groups of people emerge from the study itself. They differ demonstrably in how they respond. The patterns are interpretable and suggest different "mental realities" of people. Furthermore, the approach to uncovering these mind-sets is purely statistical. Finally, there are procedures to assign new people to the mind-sets thus discovered using a set of questions from the original study.

The method for discovering, or perhaps better, creating these mind-sets uses k-means clustering (Mucherino et al. 2009). The underlying experimental design permits the creation of 24 vignettes in precisely the mathematical structure needed for OLS (ordinary least-squares) regression (Zdaniuk 2014). We saw the use of OLS above for the data from the total panels, both for people responding to AI-generated elements and separately for people responding to human-generated elements. We did this for the total panel of 100 respondents, putting all the data together to create a dataset of 2400 rows.

The clustering comprises a more granular analysis. The 24 vignettes for each respondent are subjected to the same OLS regression. The clustering analysis (k-means) works by computing a "distance" between pairs of respondents. For most Mind Genomics studies, the "distance" is defined as the quantity (1-Pearson R) between the two matched sets of 16 coefficients. When the two respondents show parallel patterns, the Pearson R is 1, and the distance, therefore, becomes 0 (1–1 = 0). When the two respondents show inverse patterns, going in opposite directions, the Pearson R is −1, and the distance becomes 2 (1- - 1 = 2). Once the distances are defined, the clustering method puts the respondents into two and then into three groups, called clusters. In the language of Mind Genomics, these clusters are called mind-sets. The mind-sets are homogeneous and different from each other. Within a mind-sets or cluster, the "distances" between pairs of respondents are small. Across mind-sets, the averages of the mind-sets on the 16 different elements are very different from each other. Thus, the clustering juggles two criteria, bringing elements together to form a cluster or mind-sets and, at the same time, pushing their averages apart. The clustering method can be used on any of the binary-dependent variables and even on the coefficients for RT (response time) (Mucherino et al. 2009).

Clustering first assigns each respondent to one of two clusters (two mind-sets solution), and then to one of three clusters (three mind-sets solution). The OLS analysis then uses only the data from the respondents in the appropriate cluster. For this study and for the other studies in this book the clustering itself was done on the coefficients emerging when the dependent variable was R54x. Afterward, the regression equations were first estimated using R54x as the dependent variable, but the other binary transformed variables were also estimated for the separate clusters or segments. (In the language of Mind Genomics, these clusters are called mind-sets to link them to the organizing principle of mind-sets in the population, the basic focus of Mind Genomics studies.

Table 8.14 shows the data from the human-generated elements. Table 8.15 shows the comparable data from the AI-generated elements. To review, the clustering for each table was done on the total panel for the study, the dependent variable was R54x (interests), the variables used for the clustering were the coefficients. Finally,

Table 8.14 Coefficients for two and three mind-sets for human-generated elements

Human-Generated Elements Family Stress Coefficient R54	Total	MS1/2	MS22	MS1/3	MS2/3	MS33
Base (number of respondents in this group)	**100**	**41**	**59**	**30**	**36**	**34**
Strong for MS 1 of 2 – Focus on addiction						
I think addiction to bad habits, drugs, alcohol are the main reason for the stress in the family.	14	28	5	27	7	11
Our family does not have addiction-related issues.	15	25	9	24	7	17
There are many reasons other than addiction issues to have stress in the family.	14	23	8	23	11	9
Strong for MS 2 of 2 – Focus on Relationships						
I think family relationship issues are the main reason for the stress in the family.	18	8	24	9	25	17
I do not think family relationship issues are the main reason for the stress in the family.	18	11	23	13	26	15
There are many reasons other than family relationship issues to have stress in the family.	16	6	23	7	28	13
There are many reasons other than health issues to have stress in the family.	21	20	22	20	27	15
Our family does not have family relationships issues.	17	10	21	13	24	13
Not strong for either of the two mind-sets						
I think economic-related issues are the main reason for the stress in the family.	16	12	19	10	13	26
There are many reasons other than health issues to have stress in the family.	16	13	17	15	22	10
I think health-related issues are the main reason for the stress in the family.	14	12	16	14	21	7
Our family does not have economic issues.	14	9	16	6	8	25
I don't think health-related issues are the main reason for the stress in the family.	16	17	15	16	24	9
There are many reasons other than economic issues to have stress in the family.	17	17	15	15	6	27
I do not think the economic-related issues are the main reason for the stress in the family.	11	7	14	2	6	25
I do not think addiction to bad habits, drugs, alcohol are the main reason for the stress in the family.	10	20	3	15	4	11

we clustered twice, first to generate two clusters or mind-set, second to generate three clusters or mind-sets. To keep the analysis simple, we focus only on the two-cluster or two mind-set solution for each study, when we do the AI summarization of the mind-sets.

The pair of two mind-set solutions, human-generated elements versus AI-generated elements, show radically different organizing principles. The

Table 8.15 Coefficients for two and three mind-sets for AI-generated elements

AI-Generated Elements Family Stress Coefficient R54	**Hu-man**	**/2**	**/2 MS1**	**MS2**	**/3**	**/3 MS1**	**/3 MS2**	**MS3**
Base (number of respondents in this group)	**1 00**		**4 4**	**6**	**5**	**3 5**	**1**	**3 3 4**
Strong for MS1/2 = Focus on health issues								
Health Issues are Both Cause and Effect of Family Stress	2 2		3 0	6	1	3 3		2 8
Health Issues are the Primary Source of Family Stress	2 0		2 7	6	1	2 9		2 7
Health Issues can be Stress Depending on Individual and Family Circumstances	2 2		2 6	9	1	2 6		3 2
Health Issues are Important but Not the Only Family Stressors	1 6		2 3	0	1	2 7		2 2
Strong for MS2/2 – Economic issues								
Economic Issues are Just One of Many Factors of Family Stress	2 4		1 0	5	3	6	4	3 4
Economic Stress is a Symptom of Deeper Issues in Family	2 5		1 4	3	3	1 1	4	3 0
Economic Issues are Stress depending on the Context or Stage of Life	2 2		1 2	0	3	4	6	2 9
Economic Issues are the Primary Source of Family Stress	2 2		1 2	9	2	8	9	2 9
Not strong for either of the two mind-sets								
Family Relationships are the Primary Source of Stress	1 9		2 0	8	1	2 0	6	1 2 0
Family Relationships can be Stress Depending on the Family's Situation and Dynamics	1 6		1 6	7	1	1 9	0	1 2 0
Family Relationships are Just One Aspect of Stress	1 5		1 2	7	1	1 4	0	1 1 9
Family Relationships are Both Cause and Effect of Family Stress	1 5		1 6	3	1	1 9		1 68
Addictions cause Family Stress Depending on the Severity and Impact of the Addiction	1 4		2 0	0	1	1 8	3	2 2
Addictions are One of Many Family Stressors	1 2		1 6		9	1 3	0	2 3
Addictions are Both Cause and Consequence of Family Stress	1 3		2 0		7	2 2	7	1 0
Addictions are the Primary Source of Family Stress	1 1		2 0		4	1 9	0	2 0

elements generated by humans drive a two-cluster solution focusing on addiction versus family. The elements generated by AI drive a two-cluster solution focus on health versus economics.

It is clear just by looking at Tables 8.14 and 8.15 that the elements created by human beings generated higher coefficients in general than did the elements created by AI. Our first conclusion, therefore, is that the human elements generate stronger responses. However, there is a deeper pattern to the data, which may cast a shadow on the clearly stronger performance of the elements across mind-sets. This is the conclusion reached after computing the IDT, the index of divergent thought (Runco and Acar 2019; Abdulla Alabbasi et al. 2021; Thakral et al. 2021).

The IDT was developed to understand how well the Mind Genomics effort was used to learn about people's thoughts. The original notion was that it was better to have high coefficients rather than low coefficients. The IDT was suggested as a computational formula, which showed the dynamic tension between a study generating a lot of high positive coefficients for two or three mind-sets, versus a study which showed that some elements performed strongly for one mind-set, and poorly for another.

The computational formula is shown in Table 8.16 (top for human-generated elements, bottom of table for AI-generated elements). At first it was thought that a "good study" was one which showed high coefficients for the total and for the mind-sets. High coefficients, it was reasoned, seem to "tap into the mind of the

Table 8.16 Calculation of the IDT (Index of Divergent Thought) for elements generated by humans versus elements generated by AI

IDT CALULATION	Total	MS1/2	MS2/2	M1/3	MS2/3	MS3/3
Family Stress Human-Generated Elements						
Sum Total of Squares of All Positive Coefficients	3921	4224	4546	3969	5431	4574
Average of All Positive Coefficients (sum coefficients/base)	39	103	77	132	151	135
Number of respondents in the column (in the subgroup)	100	41	59	30	36	34
Weight = Proportion of the respondents in the subgroup	0.33	0.14	0.2	0.1	0.12	0.11
Weighted Total (Weight x Sum Total of Squares of All Positive Coefficients)	1294	591	909	397	652	503
Final Score = sum of weighted total	4346	66				
IDT CALULATION						
Family Stress AI-Generated Elements						
Sum Total of Squares of All Positive Coefficients	5490	5930	6345	6228	6633	8257
Average of All Positive Coefficients (sum coefficients/base)	55	135	113	178	214	243
Number of respondents in the column (in the subgroup)	100	44	56	35	31	34
Weight = Proportion of the respondents in the subgroup	0.33	0.15	0.19	0.12	0.1	0.11
Weighted Total (Weight x Sum Total of Squares of All Positive Coefficients)	1812	890	1206	747	663	908
Final Score = sum of weighted total	6226	79				

respondent." It was also, however, becoming apparent that very high IDT values corresponded to elements that performed well across all mind-sets. The basic idea for mind-sets was that the same element should perform well in one of several mind-sets and poorly in the other mind-set, NOT THAT THE ELEMENT SHOULD PERFORM WELL IN ALL MIND-SETS. Simulations suggested that IDT values around 70 corresponded to strong performing elements in some mind-sets and not in others, a finding that is more meaningful than having strong performing elements in all mind-sets.

Based upon the newly developed rule of thumb that IDT values (Moskowitz et al. 2020) of 70 or so are reflections of stronger segmentations and more dissimilar clustering, we are left with the interesting result that the IDT value for human-generated elements is 66. In contrast, the IDT value for AI-suggest elements is 79. The human-generated elements may perform less well but more powerfully demonstrate differences among mind-sets.

Table 8.16 shows the coefficients and IDT (Index of Divergent Thought) for AI and human-generated elements. The table shows the coefficients for Total Panel, three mind-set solution and two mind-set solution respectively. The table is sorted in descending order by the strong performing elements in the three-mind-set solution.

8.6 AI Summarization of the Mind-Sets

The AI system was required to create a label for the pair of two-segment solutions, based upon the meaning of the elements with a coefficient of 21 or higher, when the dependent variable was R54x (interest). Table 8.17 provides us with the AI-developed summary.

8.7 Conclusions

This chapter compares two experiments using the same set of questions or categories. The elements/answers for each category were created by researchers and by AI. The goal was to determine the nature of the elements created by the two methods. Based upon the newly developed rule of thumb that IDT values (Moskowitz et al. 2020) of 70 or so are reflections of stronger segmentations and more dissimilar clustering, we conclude the exciting result that the IDT value for AI-generated elements is 79. In contrast, the IDTIndex of Divergent Thought (IDT) value for human-suggested elements is 66. The AI-generated elements may perform better in terms of their ability to drive "interesting," but they do less well in highlighting difference among the segments. The AI performs strongly, independent of mind-set.

Table 8.17 AI-developed summary

Human-Generated elements—Mind-sets 1 of 2
Interested in:
Health Issues are the Primary Source of Family Stress; Health Issues are Important but Not the Only Family Stressors; Health Issues can be Stress Depending on Individual and Family Circumstances; Health Issues are Both Cause and Effect of Family Stress;
Create a label for this segment:
"Health Issues as a Dominant but Contextually Dependent Source of Family Stress"
Describe this segment:
This segment of the total sample in the research study recognizes that health issues are the primary source of family stress. They understand that health issues play a significant role in causing stress within the family dynamics. However, they also acknowledge that health issues are not the only sources of stress in a family. They believe that other factors can also contribute to stress within the family unit. They recognize that individuals and families may respond to health issues differently, with some finding them more stressful than others depending on individual and family circumstances. Moreover, this segment acknowledges that health issues can be both a cause and an effect of family stress. In other words, health issues can contribute to stress within the family, but the stress itself can also impact the overall health of the family members.
Human-Generated Elements—Mind-Sets 2 of 2
Interested in:
Economic Issues are the Primary Source of Family Stress; Economic Issues are Just One of Many Factors of Family Stress; Economic Issues are Stress depending on the Context or Stage of Life; Economic Stress is a Symptom of Deeper Issues in Family;
Create a label for this segment:
Economic Stressful Factors in Family Dynamics
Describe this segment:
This segment of the research study suggests that economic issues play a significant role in causing stress within families. It can be inferred that for these individuals, economic issues are the main source of stress within their families. Additionally, it is mentioned that economic issues are just one among several factors contributing to family stress, indicating that there may be other sources of stress in these families as well. The description also suggests that the impact of economic issues on family stress may vary depending on the context or stage of life—suggesting that economic stress may be more or less pronounced during different life phases or circumstances. Furthermore, it is implied that economic stress may be a symptom of deeper underlying issues within families, indicating that financial struggles may be indicative of other problems affecting the family unit.
AI-Generated Elements Mind-Sets 1 of 2
Interested in:
Health Issues are the Primary Source of Family Stress; Health Issues are Important but Not the Only Family Stressors; Health Issues can be Stress Depending on Individual and Family Circumstances; Health Issues are Both Cause and Effect of Family Stress;
Create a label for this segment:
"Health Issues as a Dominant but Contextually Dependent Source of Family Stress"

(continued)

Table 8.17 (continued)

Human-Generated elements—Mind-sets 1 of 2
Describe this segment:
This segment of the total sample in the research study recognizes that health issues are the primary source of family stress. They understand that health issues play a significant role in causing stress within the family dynamics. However, they also acknowledge that health issues are not the only sources of stress in a family. They believe that other factors can also contribute to stress within the family unit. They recognize that individuals and families may respond to health issues differently, with some finding them more stressful than others depending on individual and family circumstances. Moreover, this segment acknowledges that health issues can be both a cause and an effect of family stress. In other words, health issues can contribute to stress within the family, but the stress itself can also impact the overall health of the family members.
AI-Generated Elements Mind-Sets 2 of 2
Interested in:
Economic Issues are the Primary Source of Family Stress; Economic Issues are Just One of Many Factors of Family Stress; Economic Issues are Stress depending on the Context or Stage of Life; Economic Stress is a Symptom of Deeper Issues in Family;
Create a label for this segment:
Economic Stressful Factors in Family Dynamics
Describe this segment:
This segment of the research study suggests that economic issues play a significant role in causing stress within families. It can be inferred that for these individuals, economic issues are the main source of stress within their families. Additionally, it is mentioned that economic issues are just one among several factors contributing to family stress, indicating that there may be other sources of stress in these families as well. The description also suggests that the impact of economic issues on family stress may vary depending on the context or stage of life—suggesting that economic stress may be more or less pronounced during different life phases or circumstances. Furthermore, it is implied that economic stress may be a symptom of deeper underlying issues within families, indicating that financial struggles may be indicative of other problems affecting the family unit.

References

Abdulla Alabbasi, A. M., Reiter-Palmon, R., Sultan, Z. M., & Ayoub, A. E. A. (2021). Which Divergent Thinking Index Is More Associated With Problem Finding Ability? The Role of Flexibility and Task Nature. *Frontiers in Psychology, 12.* https://doi.org/10.3389/fpsyg.2021.671146

Chen, J.-H., Huang, C.-H., Wu, C.-F., Jonson-Reid, M., & Drake, B. (2024). The application of family stress model to investigating adolescent problematic behaviors: The moderating role of assets. *Journal of Family and Economic Issues, 45*(1), 174–183. https://doi.org/10.1007/s10834-023-09902-2

Cherry, D. B. (1989). Stress and coping in families with ill or disabled children: application of a model to pediatric therapy. *Physical & Occupational Therapy in Pediatrics*, *9*(2), 11–32. http://www.ncbi.nlm.nih.gov/pubmed/15772024

Gofman, A., & Moskowitz, H. (2010). Isomorphic permuted experimental designs and their application in conjoint analysis. *Journal of Sensory Studies, 25*(1), 127–145. https://doi.org/10.1111/j.1745-459X.2009.00258.x

Heyl, J. C. (2022). *What Is Family Stress Theory?* https://www.verywellmind.com/what-is-family-stress-theory-6831762

Hill, E. J., Allsop, D. B., LeBaron, A. B., & Bean, R. A. (2017). How do Money, Sex, and Stress Influence Marital Instability? *Journal of Financial Therapy*, *8*(1). https://doi.org/10.4148/1944-9771.1135

Hill, R. (1949). *Families under stress: Adjustment to the crises of war separation and reunion*. Harper & Brothers.

Kover, A., Papajorgji, P., & Moskowitz, H. (2022). *Applying Mind Genomics to Social Sciences*. IGI Global. https://doi.org/10.4018/978-1-7998-8409-5

Masarik, A. S., & Conger, R. D. (2017). Stress and child development: a review of the Family Stress Model. *Current Opinion in Psychology*, *13*, 85–90. https://doi.org/10.1016/j.copsyc.2016.05.008

McCubbin, H. I., & Patterson, J. M. (1983). The Family Stress Process. *Marriage & Family Review*, *6*(1–2), 7–37. https://doi.org/10.1300/J002v06n01_02

Moskowitz, H., Lowenstein, M., Moskowitz, N. L., Danny, Y., Moskowitz, P. D., & Yehoshua, D. (2020). Experimenting & learning to think critically and competently: Combining 2020 technology with student-driven research. *Psychology Journal: Research Open*, *2*(1).

Mucherino, A., Papajorgji, P., & Pardalos, P. (2009). Clustering by k-means. In *Data Mining in Agriculture* (pp. 47–82).

Neppl, T. K., Senia, J. M., & Donnellan, M. B. (2016). Effects of economic hardship: Testing the family stress model over time. *Journal of Family Psychology*, *30*(1), 12–21. https://doi.org/10.1037/fam0000168

Runco, M. A., & Acar, S. (2019). Divergent Thinking. In *The Cambridge Handbook of Creativity* (pp. 224–254). Cambridge University Press. https://doi.org/10.1017/9781316979839.013

Thakral, P. P., Yang, A. C., Addis, D. R., & Schacter, D. L. (2021). Divergent thinking and constructing future events: dissociating old from new ideas. *Memory*, *29*(6), 729–743. https://doi.org/10.1080/09658211.2021.1940205

Zdaniuk, B. (2014). Ordinary Least-Squares (OLS) Model. In *Encyclopedia of Quality of Life and Well-Being Research* (2014. Edit, pp. 4515–4517). Springer Netherlands. https://doi.org/10.1007/978-94-007-0753-5_2008

Zietz, S., Lansford, J. E., Liu, Q., Long, Q., Oburu, P., Pastorelli, C., Sorbring, E., Skinner, A. T., Steinberg, L., Tapanya, S., Tirado, L. M. U., Yotanyamaneewong, S., Alampay, L. P., Al-Hassan, S. M., Bacchini, D., Bornstein, M. H., Chang, L., Deater-Deckard, K., Di Giunta, L., … Gurdal, S. (2022). A longitudinal examination of the family stress model of economic hardship in seven countries. *Children and Youth Services Review*, *143*, 106661. https://doi.org/10.1016/j.childyouth.2022.106661

Chapter 9
Experiment in Environmental Policies

9.1 Introduction

The issue of climate change is one of the most critical issues of our time, and it has been the subject of much discussion in the political area as well as in the scientific community. The international scientific community has started paying much attention to this issue. As a result, the first international convention that paved the way for recognizing the problem and designing well-crafted policies was the Kyoto Protocol in 1997.

The Kyoto Protocol represents a significant milestone in climate policy development. For the first time, negotiators have made an effort to spell out goals for cutting emissions in the first half of the twenty-first century (Manne and Richels 2000).

Reducing greenhouse gas emissions was one of the Kyoto Protocol's primary objectives in the fight against global climate change. It established legally mandated goals for wealthy nations to cut their emissions between 2008 and 2012 by an average of 5% below 1990 levels (United Nations 1997). To assist nations in achieving these reductions at a reasonable cost, the protocol incorporated market-based mechanisms, including emissions trading, the Clean Development Mechanism (CDM), and Joint Implementation (JI).

Following its global signature in December 1997, lawmakers debated whether to ratify and execute the "Kyoto Protocol." The treaty, widely regarded as the first significant step toward reducing global warming, mandates that all developed countries control their emissions at predetermined levels. Those targets cover the "budget period" of 2008–2012, and as part of the protocol, countries will also agree on restrictions for subsequent budget periods. Notwithstanding mounting public pressure to address global warming, lawmakers will consider the expense of ratifying the Kyoto Agreement before taking action (VICTOR 2011).

The Kyoto Agreement aided the mainstreaming of climate change topics in research. Kim (2021) employed the number of patents for technology aimed at

P. Papajorgji, H. Moskowitz, *The Mind of Everyday*,
https://doi.org/10.1007/978-3-031-78078-3_9

mitigating climate change as a gauge of innovation acquired from 54 nations between 1990 and 2015 to ascertain whether technological innovation is successful in cutting greenhouse gas emissions and has a strong association with the Kyoto Protocol.

The Paris Agreement is the most important international climate pact, second to the Kyoto Protocol, and it was approved during the December 2015 COP21 conference in Paris. The Kyoto Protocol concentrated on lowering greenhouse gas emissions by setting legally binding reduction targets, especially for wealthy countries. At the same time, the Paris Agreement takes a more comprehensive approach with multiple important debate points and conclusions (Skjærseth et al. 2021).

Paris Agreement was a shift from the approach embraced by the Kyoto Protocol (Paris Agreement 2024). First, it recognized that climate change is a global issue that needs a global common approach. Second, it recognized the importance of non-state actors, implicating all nations in cutting global emissions and emphasizing the need for international cooperation to overcome climate change hurdles successfully (Huang and Zhai 2021).

As with any other important issue, there are voices pro and anti the relevance and success of the Paris Agreement. Raiser et al. (2020) makes a detailed and well-informed analysis to answer the Paris Agreement‘s implementation question. Clémençon (2016) states that the world is better off with a non-perfect agreement than nothing at all. Mulatu et al. (2024) point out that the objectives of the Paris Agreement can be implemented with success only if a global and transparent coalition is assembled to address this crucial problem.

The preservation of biodiversity is becoming a constant discussion in the current scientific literature. This issue is now more widely acknowledged as a significant issue in the environmental policies of numerous governments worldwide. This change is the result of a greater awareness of the vital role that biodiversity plays in preserving ecosystem services, which are necessary for regulating climate, ensuring economic stability, and promoting human well-being (Pimm 2021; Kilpatrick et al. 2017). But, Wood et al. (2014) state there is little causal evidence to support biodiversity conservation as a win–win strategy for preserving ecosystem health and safeguarding public health. Wood et al. (2017) clearly state that human infectious disease burdens decrease with urbanization but not with biodiversity.

To promote biodiversity conservation, we must gain a deeper comprehension of the connections between ecosystem services and various dimensions of biodiversity (Wu et al. 2022).

One relevant point needs to be made on the role of technology in reaching the objectives established by the Kyoto and Paris Agreements. Clean technologies are seen as the best approach to reaching the ambitious goals of the Kyoto and Paris Agreements. According to our estimates, the world does not achieve climate goals in 2100 without "green" technologies (Alestra et al. 2023). Jurkus et al. (2022) highlight the key developments necessary to meet the UN's Sustainable Development Goals by 2030.

The world's long-term sustainability is severely hampered by global environmental degradation (Awewomom et al. 2024; Ukaogo et al. 2020). The repercussions of pollution on society necessitate a multidimensional response that includes international collaboration, public awareness campaigns, technological advancements, and regulatory measures. In addition to reducing the negative consequences of pollution, effective policies and practices can enhance social cohesion, public health, and economic stability (Abubakar et al. 2022; Hajam et al. 2023).

Considering the significant problems caused by inappropriate waste management practices, waste management strategies should be aligned with circular economic models that consider the economy, environment, and efficiency (Kumar et al. 2023).

Targeting energy efficiency as a road map for mitigating carbon emissions, restricting energy use, enhancing building energy performance, and lowering energy consumption to achieve sustainable buildings has garnered more attention in recent years (Hafez et al. 2023).

Studies on waste, resource, and energy management are common since sustainable growth depends on how well they are coordinated (Wang et al. 2022).

In order to minimize the detrimental effects of human activity on the environment, long-term environmental sustainability can be achieved through sustainable resource usage and energy efficiency (Vedanarayanan et al. 2022).

Based on the literature review presented above, the topics selected for the Mind Genomics experiments are public perceptions on:

- Climate change mitigation and adaptation.
- Biodiversity conservation.
- Pollution control and waste management.
- Sustainable resource use and energy efficiency.

9.2 Mind Genomics Experiments

This chapter examines the results of two approaches to the same problem—public views of government environmental policies. While one strategy relies on the knowledge of a human researcher, the other will use recommendations from an artificial intelligence (AI) system. The goal is to determine whether using AI-based expertise instead of human expertise makes a difference; the outcomes will be examined individually. The methodology used in both studies is mind genomics (Gofman and Moskowitz 2010; Moskowitz et al. 2006; Milutinovic and Salom 2016; Gofman and Moskowitz 2012).

The starting point of any Mind Genomics-based research project is analyzing the problem under inquiry to identify the most critical issues that meet the study's objectives (Todri et al. 2021; Ilollari et al. 2020). Since the concept of silos is the same in both studies, the same government environmental challenges are examined in both. Instead, the responses for every silo will differ; for example, in one study, CHATGPT 3.5 (OpenAI. 2023) will supply the answers, while a human researcher will provide the answers in the second study. Every study will undergo Mind

Genomics analysis, and the outcomes will be scrutinized to ascertain if employing an AI-driven instrument to generate the solutions makes a difference.

Both studies use the same system, https://www.bimileap.com/, as presented in Chap. 6. Both studies are designed using a 4 × 4 Mind Genomics model. Each study is evaluated using 2400 observations. Special care was shown to have similar participant distribution to ensure the data were analyzed consistently.

At the outset, participants get the following information outlining the objectives of the study:

> ***Environmental Governmental Policies involves analyzing various critical issues that are central to understanding the effectiveness, impact, and challenges of these policies. Mismanagement in governmental policies creates important economic and health issues. We would like to have your opinion to understand and address these issues.***

Table 9.1 lists the silos and the matching responses for every silo used in the human-based experiment.

Table 9.2 lists the silos and the matching responses for each silo in the AI-based experiment.

Both experiments use a five-level Likert scale to evaluate vignettes as follows. Level 1 means **Unlikely** and level 5 means **Likely.**

1 = Not interested ... AND ....won't happen
2 = Not interested ... BUT .. It could happen
3 = Can't honestly say
4 = Looks interesting .. BUT...won't happen
5 = Looks interesting .. AND .. It could happen

A balanced distribution of participant characteristics, including age, education level, and gender, was ensured with extra care.

Following participant data collection, all computations will be performed by Bimileap, and the findings will be presented in an Excel file with many sheets, as demonstrated in Chap. 6.

9.3 Comparing Results of Both Experiments: What We Found

As shown in Fig. 9.1, the elements generated by AI differed from those generated by researchers. In this study, the total of silos in the experiment designed by AI scored higher values than the one designed by humans. Although we obtained different and higher values from the AI-based experiment, comparing general patterns emerging from the two studies on the same topic is appropriate. There are several ways to compare the patterns of responses generated by the two experiments, ranging from the patterns of ratings, the relations between the presence/absence of elements versus the response to the vignette (modeling), the interpretation of strong

Table 9.1 Silos and elements defined by researchers

Question A: Do you think climate change mitigation and adaptation are the most pressing issues in government environmental policy?	
A1	Government should consider climate change mitigation and adaptation as one of the most important issues
A2	Governments should give more importance to other issues such as industrialization, job creation, and poverty reduction
A3	I lack sufficient information to provide a conclusive response about climate change mitigation and adaptation
A4	My life is full of activity, and I don't have time to think about climate change
Question B: Do you think biodiversity conservation is a significant concern of the government's environmental policy?	
B1	Biodiversity conservation should be considered as a significant component of environmental policies
B2	Biodiversity conservation is a long-term issue, so the focus must be on human well-being, resource extraction, agriculture, and urban development
B3	I am not confident speaking about biodiversity conservation because my knowledge is insufficient
B4	There are more critical issues than biodiversity that the government should pay immediate attention
Question C: Do you think pollution control and waste management are crucial for protecting human health and the environment?	
C1	Pollution is a major public health issue, leading to diseases and poor waste management contributes to pollution
C2	While important, pollution control and waste management need to be balanced with economic considerations
C3	Immediate human needs such as poverty alleviation, education, and healthcare often take precedence over pollution control and waste management
C4	Market forces and technological advancements will naturally lead to better pollution control and waste management solutions without the need for stringent regulations
Question D: Do you think sustainable resource use and energy efficiency are the focus of the government's environmental policy?	
D1	Sustainable resource use and energy efficiency should be and often are key focuses of government environmental policy
D2	While sustainable resource use and energy efficiency are important, they are often balanced with economic and industrial interests
D3	While sustainable resource use and energy efficiency are recognized as important, the immediate focus might be on more pressing developmental issues
D4	Innovation and market forces, rather than government policies, are more effective in driving sustainability and efficiency

Table 9.2 Silos and the corresponding elements generated by AI

Question A: Do you think climate change mitigation and adaptation are the most pressing issues in government environmental policy?	
A1	Climate change mitigation and adaptation are the most pressing issues as it poses a significant threat to ecosystems, human health, and the economy
A2	While climate change is important, we must balance environmental concerns with economic realities
A3	For many countries, immediate issues such as poverty, access to clean water, and basic healthcare are more pressing than climate change mitigation
A4	The extent of human impact on climate change is overstated, and that natural climate variability plays a significant role
Question B: Do you think biodiversity conservation is a significant concern of the government's environmental policy?	
B1	Biodiversity conservation is a critical concern and should be a top priority in government environmental policy
B2	Biodiversity is important, but not at the cost of economic growth and job creation
B3	Biodiversity conservation should be addressed, but in a way that is practical and integrates with other environmental and social goals
B4	The emphasis on biodiversity conservation is exaggerated, nature is resilient and capable of adapting to changes
Question C: Do you think pollution control and waste management are crucial for protecting human health and the environment?	
C1	Pollution is a major public health issue, leading to diseases and poor waste management contributes to pollution
C2	While important, pollution control and waste management need to be balanced with economic considerations
C3	Immediate human needs such as poverty alleviation, education, and healthcare often take precedence over pollution control and waste management
C4	Market forces and technological advancements will naturally lead to better pollution control and waste management solutions without the need for stringent regulations
Question D: Do you think sustainable resource use and energy efficiency are the focus of the government's environmental policy?	
D1	Sustainable resource use and energy efficiency should be and often are key focuses of government environmental policy
D2	While sustainable resource use and energy efficiency are important, they are often balanced with economic and industrial interests
D3	While sustainable resource use and energy efficiency are recognized as important, the immediate focus might be on more pressing developmental issues
D4	Innovation and market forces, rather than government policies, are more effective in driving sustainability and efficiency

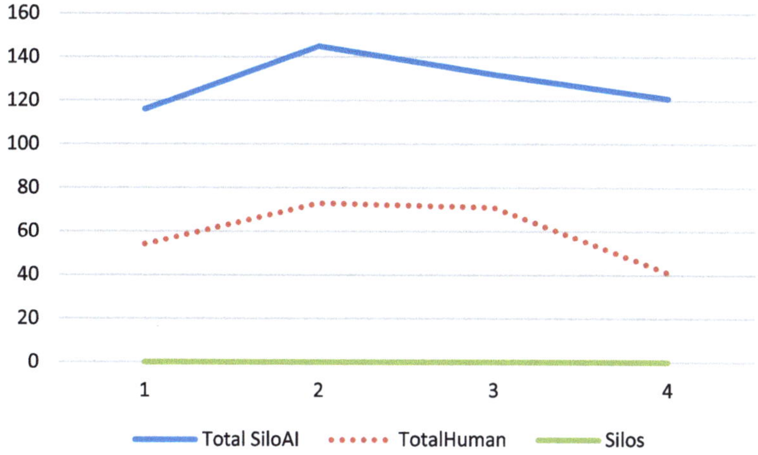

Fig. 9.1 Total silos per experiment

performing elements (viz., interpretation and AI-based summarization), and finally the division of the respondents into mind-sets by clustering and the nature of these clusters.

The Mind Genomics platform performed the necessary computations after gathering all of the participant data (see Chap. 6). Once more, using clustering based on the coefficients of the 16 elements (question answers), two and three mind-sets were extracted. The coefficients resulting from the cluster analysis, which only took into account the 16 items that participants submitted, are displayed in Tables 9.3 for the human-based experiment and 9.4 for the AI-based experiment. These elements were clustered into two and then three mind-sets, respectively.

Analyzing the data of Table 9.3, the two and three mind-sets for the human-based study, we notice that the highest-scoring value is of 26 for the two mind-set solution and 34 for the three mind-set solution. Thus, clusters in the three mind-set solution have higher scoring elements.

Table 9.4 shows the two and three mind-set values for the AI-based experiment.

Analyzing the data of Table 9.4, the two and three mind-sets for the AI-based study, we notice that the highest-scoring value is of 35 for the two mind-set solution and 30 for the three mind-set solution. Thus, clusters in the two mind-set solution have higher scoring elements.

In order to facilitate the analyses, a transformation from a five-scale evaluation to a binary evaluation is carried out before beginning the analysis of the data from the two experiments, as explained in Chap. 6. The underlying statistical analysis produces new binary variables. The change is shown in Table 9.5. The original 5-point scale itself does not give us the required format to perform some of the studies we require, which is why we are transforming it (viz., modeling and clustering).

The binary variable R54x, produced when ratings 5 and 4 are converted to 100 while ratings 2, 3, and 1 are converted to 0, is the subject of this study. To this newly created binary dependent variable (BDV), a vanishingly small random value (<10–5)

Table 9.3 The two and three mind-set solution for the human-based experiment

	Human generated elements	**Total**		**Mindset 1 of 2**	**Mindset 2 of 2**		**Mindset 1 of 3**	**Mindset 2 of 3**	**Mindset 3 of 3**
	Base Size	100		53	47		25	35	40
	Question A: Do you think climate change mitigation and adaptation are the most pressing issues in government environmental policy?								
A1	Government should consider climate change mitigation andadaptation as one of the most important issues.	12		20	2		19		22
A2	Governments should give more importance to other issues such as industrialization, job creation, and poverty reduction.	14		24	2		16	4	21
A3	I lack sufficient information to provide a conclusive response about climate change mitigation and adaptation.	13		19	5		10	4	21
A4	My life is full of activity, and I don't have time to think about climate change	15		22	9		19	6	20
	Question B: Do you think biodiversity conservation is a significant concern of the government's environmental policy?								
B1	Biodiversity conservation should be considered as a significant component of environmental policies.	19		18	19			21	26
B2	Biodiversity conservation is a long	16		11	23			25	19

Table 9.4 The two and three mind-set solution for the AI-based experiment

	AI Generated Elements	**Total**		**Mindset 1 of 2**	**Mindset 2 of 2**		**Mindset 1 of 3**	**Mindset 2 of 3**	**Mindset 3 of 3**
	Base Size	100		55	45		48	28	24
	Question A: Do you think climate change mitigation and adaptation are the most pressing issues in government environmental policy?								
A1	Climate change mitigation and adaptation are the most pressing issues as it poses a significant threat tœcosystems, human health, and the economy.	14		19	7		18	14	3
A2	While climate change is important, we must balance environmental concerns with economic realities.	13		17	10		17	17	2
A3	For many countries, immediate issues such as poverty, access to clean water, and basic healthcare are more pressing than climate change mitigation.	12		19	2		21	12	
A4	The extent of human impact on climate change is overstated, and that	15		21	6		22	16	

is introduced. The value is inserted as a precautionary measure to guarantee that R54x has some minimal variability and can be used in OLS (ordinary least-squares regression).

The first "morphological" study looks at the percentages of rating points assigned to the vignettes. The primary subject of interest is the difference between the rating distributions of elements chosen by AI and those chosen by humans. Again, the only details that are known about the various parts are what their sources claim. Responses originating from human-sourced ideas get ratings of "5," which are slightly lower than responses originating from ideas drawn from other sources; 26 for AI-based study versus 23 for the human-based study as shown in Table 9.6. This fact points

Table 9.5 Transformation of 5-rating to a 2-rating scale

	The original points on the rating scale				
	1	2	3	4	5
Left side of scale: interest me	N	N	DK	Y	Y
Tight side of scale: Happen	N	Y	DK	N	Y
New binary variables					
R1x	100	0	0	0	0
R2x	0	100	0	0	0
R3x	0	0	100	0	0
R4x	0	0	0	100	0
R5x	0	0	0	0	100
R54x	0	0	0	100	100
R52x	0	100	0	0	100
R41x	100	0	0	100	0
R21x	100	100	0	0	0

Table 9.6 Frequency of scale point per vignette

Environment	Original scale points Numbers in body of table are percent of times the rating scale was assigned to a vignette				
	1	2	3	4	5	Total
Source						
Total	12	12	24	28	24	100
AI	10	11	26	27	26	100
Human	13	14	22	29	23	100

to a slight divergence in thought between the two sources. As seen below, this discovery will be noteworthy and have fascinating implications.

In the next step of our examination, we delve into the performance of the elements. R54x, a significant variable, serves as a measure of how well these elements perform. R54x is the union of the two scale points, 4 and 5, with the property "it looks interesting." The technique that facilitates the connection between the solution and the elements is known as ordinary least-squares regression (Zdaniuk 2014), or OLS for short.

The study's setup enables the researcher to create equations by regression because each respondent's underlying experimental design guarantees that the 16 elements are presented in a way that makes them statistically independent of each other and has the property that the 24 vignettes each respondent evaluates are different from the 24 vignettes evaluated by the other respondent.

The analysis for the total panel for each study creates this simple equation: R54x = k1A1 + k2A2 … k16A16.

Each element's driving power is shown by the coefficients. These coefficients, k1–k16, achieve statistical significance when they approach or surpass 10–12, according to statistical assessments. The focus is on items that have relevance coefficients of 16 or higher and importance coefficients of 21 or higher. Experience determines the coefficients' cutoff level. It is stricter than a straightforward analysis of the coefficient's T value as determined by the regression model's analyses of variance.

Table 9.7 shows the coefficients of the 16 elements from OLS regression, showing the results from human-generated elements.

Table 9.7 displays the performance of human-generated items. Finally, one factor (coefficient of 20) turns out to be a particularly good performer: "**Immediate**

Table 9.7 Performance of element R54x for the human-based experiment

	Performance of HUMAN-generated elements (Dependent variable = R54x	
	Base Size	100
C3	Immediate human needs such as poverty alleviation, education, and healthcare often take precedence over pollution control and waste management.	20
B1	Biodiversity conservation should be considered as a significant component of environmental policies.	19
B3	I am not confident speaking about biodiversity conservation because my knowledge is insufficient.	19
B4	There are more critical issues than biodiversity that the government should pay immediate attention.	19
C4	Market forces and technological advancements will naturally lead to better pollution control and waste management solutions without the need for stringent regulations.	19
C1	Pollution is a major public health issue, leading to diseases and poor waste management contributes to pollution.	17
B2	Biodiversity conservation is a long-term issue, so the focus must be on human well-being, resource extraction, agriculture, and urban development.	16
A4	My life is full of activity, and I don't have time to think about climate change.	15
C2	While important, pollution control and waste management need to be balanced with economic considerations.	15
A2	Governments should give more importance to other issues such as industrialization, job creation, and poverty reduction.	14
A3	I lack sufficient information to provide a conclusive response about climate change mitigation and adaptation.	13
A1	Government should consider climate change mitigation and adaptation as one of the most important issues.	12
D1	Sustainable resource use and energy efficiency should be and often are key focuses of government environmental policy.	11
D3	While sustainable resource use and energy efficiency are recognized as important, the immediate focus might be on more pressing developmental issues.	11
D4	Innovation and market forces, rather than government policies, are more effective in driving sustainability and efficiency.	10
D2	While sustainable resource use and energy efficiency are important, they are often balanced with economic and industrial interests.	9

human needs such as poverty alleviation, education, and healthcare often take precedence over pollution control and waste management." The reason behind this specific element's high score is unclear.

"**While sustainable resource use and energy efficiency are important, they are often balanced with economic and industrial interests**" is the element with the lowest performance, scoring a coefficient of 9. Even though every element is positive and at least 9, the small range of coefficients is unexpected. It is interesting to see whether this tight range will widen when we address values in experiments that are created by AI-based study.

Table 9.8 shows the coefficients of the 16 elements from OLS regression, showing the results from AI-generated elements. Only one element, "**Biodiversity conservation is a critical concern and should be a top priority in government environmental policy**"has a score of 20, which is the highest score.

Only 7 elements score between 16 and 20 in the human-based study versus 8 in the AI-based study. Thus, both studies have the same highest score and the similar number of scoring elements in the range 16–20.

As a summary of what this first exploration tells us, we can say that, for both studies and the entire panel, our coefficients perform low to moderately. None of the elements have incredibly low scores, nor do any of the elements score remarkably well, with coefficients of 30 or higher coefficients that, overall, are highly noteworthy and deserving of notice. However, our initial reaction should be that AI produces things that are reasonable, if not exceptionally powerful.

9.4 AI Summarization of Strong Elements

An AI-based summarizer utilizing Chap GPT4 (OpenAI 2023) was employed as a component of the Mind Genomics platform to investigate various facets of high-performing parts with a score of 21 or above. None of studies have elements scoring 21 or higher. Thus, the summarizer cannot be used in these studies.

9.5 Beyond the Single Dependent Variable to Considering the Pattern for All Variables

By using the Mind Genomics technique, the researcher can assess how each component contributes differently to a variety of binary dependent variables. With the exception of AI-generated items, which perform relatively better than human-generated ones (the sum of AI-generated items is 248 and the mean is 15.5 while the sum of human-generated items is 239 and the mean 14.9), we have a sense of how closely different elements perform in terms of R54x because we have only looked at one dependent variable thus far: R54x.

Table 9.8 Performance of element R54x for the AI-based experiment

	Performance of AI-generated elements (Dependent variable = R54x)	**Total**
	Base Size	100
B1	Biodiversity conservation is a critical concern and should be a top priority in government environmental policy.	20
B2	Biodiversity is important, but not at the cost of economic growth and job creation.	19
B4	The emphasis on biodiversity conservation is exaggerated, nature is resilient and capable of adapting to changes.	17
C2	While important, pollution control and waste management need to be balanced with economic considerations.	17
C3	Immediate human needs such as poverty alleviation, education, and healthcare often take precedence over pollution control and waste management.	17
C4	Market forces and technological advancements will naturally lead to better pollution control and waste management solutions without the need for stringent regulations.	16
D1	Sustainable resource use and energy efficiency should be and often are key focuses of government environmental policy.	16
D4	Innovation and market forces, rather than government policies, are more effective in driving sustainability and efficiency.	16
A4	The extent of human impact on climate change is overstated, and that natural climate variability plays a significant role.	15
B3	Biodiversity conservation should be addressed, but in a way that is practical and integrates with other environmental and social goals.	15
D2	While sustainable resource use and energy efficiency are important, they are often balanced with economic and industrial interests.	15
A1	Climate change mitigation and adaptation are the most pressing issues as it poses a significant threat to ecosystems, human health, and the economy.	14
C1	Pollution is a major public health issue, leading to diseases and poor waste management contributes to pollution.	14
A2	While climate change is important, we must balance environmental concerns with economic realities.	13
A3	For many countries, immediate issues such as poverty, access to clean water, and basic healthcare are more pressing than climate change mitigation.	12
D3	While sustainable resource use and energy efficiency are recognized as important, the immediate focus might be on more pressing developmental issues.	12

The study that follows looks at the spectrum of coefficients for each of the binary dependent variables shown in Table 9.5. The analysis and modeling were identical.

To gain more insight into the differences between AI-generated and human-generated elements, we use OLS regression to link the recently built BDVs and response time to the presence or absence of elements. We have already completed that analysis. Table 9.10 shows the coefficients for the AI-generated elements. Table 9.11 shows the coefficients for the elements that were created by humans. Tables show darkened cells that draw attention to the underlying patterns.

The shading criteria that are operationally specified are as follows:

When the coefficient is 21 or greater, R54x is shaded.
When the response time is 1.2 or more, the RT is shaded.

In Tables 9.9 and 9.10 we have considered only R54x and R21x because scale has only one variable and R52 and R41 not meaningful.

9.6 Moving from the Total Panel to the Mind-Sets, and to the IDT (Index of Divergent Thought)

One distinctive feature of mind genomics is its focus on specific concerns rather than broad subjects, and on mind-sets developed from the granularity of the data. Mind Genomics investigates the pattern of responses to specific themes, such as those discussed here, starting from the ground up. Using this little "slice" of reality, Mind Genomics aims to identify interpretable and potentially meaningful groups of people displaying different responses to the same stimuli. These ideas originate from the actual investigation. There is a definite difference between their answers.

The patterns are interpretable and suggest different people's "mental realities." Furthermore, only statistical methods are employed to identify these mind-sets. Finally, techniques have been developed for categorizing recently hired individuals into the aforementioned mind-sets; these techniques utilize a number of questions from the initial study. This technique is called VPI (the View Point Identifier) (Zemel et al. 2019).

k-means clustering is the technique for identifying, or maybe more effectively, generating these mind-sets (Mucherino et al. 2009). Twenty-four vignettes can be created in precisely the mathematical structure required for ordinary least-squares (OLS) regression thanks to the underlying experimental design. As we saw above, OLS was applied to the complete panel data for both the responses of individuals to AI-generated elements and the responses of individuals to human-generated elements separately. We completed this for each of the 100 respondents in the panel, combining all the information into a dataset with 2400 rows.

Following clustering, an in-depth examination is carried out. The 24 vignettes from each respondent are subjected to the same OLS regression. "Distance" between each pair of responders is calculated using the k-means clustering algorithm. For

Table 9.9 Coefficients for binary-dependent variables, and RT for AI-generated elements

Elements	Environment AI	R54X	R21X	RT
A1	Climate change mitigation and adaptation are the most pressing issues as it poses a significant threat to ecosystems, human health, and the economy.	14	7	1.6
A2	While climate change is important, we must balance environmental concerns with economic realities.	13	8	1.3
A3	For many countries, immediate issues such as poverty, access to clean water, and basic healthcare are more pressing than climate change mitigation.	12	8	1.1
A4	The extent of human impact on climate change is overstated, and that natural climate variability plays a significant role.	14	11	1.3
B1	Biodiversity conservation is a critical concern and should be a top priority in government environmental policy.	20	3	1.4
B2	Biodiversity is important, but not at the cost of economic growth and job creation.	19	4	1.3
B3	Biodiversity conservation should be addressed, but in a way that is practical and integrates with other environmental and social goals.	15	6	1.3
B4	The emphasis on biodiversity conservation is exaggerated, nature is resilient and capable of adapting to changes.	17	6	1.4
C1	Pollution is a major public health issue, leading to diseases and poor waste management contributes to pollution.	14	8	1.1
C2	While important, pollution control and waste management need to be balanced with economic considerations.	17	5	1.2
C3	Immediate human needs such as poverty alleviation, education, and healthcare often take precedence over pollution control and waste management.	17	7	1.2
C4	Market forces and technological advancements will naturally lead to better pollution control and waste management solutions without the need for stringent regulations.	16	7	1.1
D1	Sustainable resource use and energy efficiency should be and often are key focuses of government environmental policy.	16	4	0.9
D2	While sustainable resource use and energy efficiency are important, they are often balanced with economic and industrial interests.	15	4	1
D3	While sustainable resource use and energy efficiency are recognized as important, the immediate focus might be on more pressing developmental issues.	12	6	0.9
D4	Innovation and market forces, rather than government policies, are more effective in driving sustainability and efficiency.	16	3	1

Table 9.10 Coefficients for binary-dependent variables, and RT for human-generated elements

Elements	Environment Human	R54X	R21X	RT
A1	Government should consider climate change mitigation and adaptation as one of the most important issues.	12	10	1.1
A2	Governments should give more importance to other issues such as industrialization, job creation, and poverty reduction.	14	10	1.3
A3	I lack sufficient information to provide a conclusive response about climate change mitigation and adaptation.	13	10	1
A4	My life is full of activity, and I don't have time to think about climate change.	15	9	1.3
B1	Biodiversity conservation should be considered as a significant component of environmental policies.	18	8	1.4
B2	Biodiversity conservation is a long-term issue, so the focus must be on human well-being, resource extraction, agriculture, and urban development.	16	8	1.3
B3	I am not confident speaking about biodiversity conservation because my knowledge is insufficient.	19	8	1.4
B4	There are more critical issues than biodiversity that the government should pay immediate attention.	19	8	1.3
C1	Pollution is a major public health issue, leading to diseases and poor waste management contributes to pollution.	17	8	1.2
C2	While important, pollution control and waste management need to be balanced with economic considerations.	15	12	1.2
C3	Immediate human needs such as poverty alleviation, education, and healthcare often take precedence over pollution control and waste management.	20	7	1.2
C4	Market forces and technological advancements will naturally lead to better pollution control and waste management solutions without the need for stringent regulations.	19	5	1.3
D1	Sustainable resource use and energy efficiency should be and often are key focuses of government environmental policy.	11	4	0.9
D2	While sustainable resource use and energy efficiency are important, they are often balanced with economic and industrial interests.	9	4	1
D3	While sustainable resource use and energy efficiency are recognized as important, the immediate focus might be on more pressing developmental issues.	11	6	1.1
D4	Innovation and market forces, rather than government policies, are more effective in driving sustainability and efficiency.	10	7	1.2

most Mind Genomics studies, "distance" is defined as the amount (1-Pearson R) separating the two matched sets of 16 coefficients.

If both responders show comparable patterns, the Pearson R is 1 and the distance is equal to 0 (1–1 = 0). When the two respondents move in opposite directions and show inverse patterns, the distance increases to 2 (1– − 1 = 2). In this case, the Pearson R value is −1. Once the distances are known, the clustering process splits the responses into two and then three groups, or clusters. In the jargon of mind genomics, these clusters are called mind-sets. The ideas are consistent and separate from one another. Tiny "distances" existed between respondent pairs within a mind-set or cluster.

Mind-sets differ significantly in the averages of their perceptions on the 16 elements. Consequently, the clustering strikes a balance between two requirements: it aggregates elements to form a cluster or thinking while maintaining the distinction between their averages. The response time and RT coefficients are two examples of binary dependent variables to which the clustering method can be applied (Mucherino et al. 2009).

Following a two mind-set solution, clustering allocates each respondent to one of two clusters, and subsequently to one of three clusters (three mind-set solution). Using only the data from the relevant cluster or mind-set, it is simple to perform the OLS regression on the binary dependent variable of choice (in this case, R54X, sounds interesting).

Table 9.11 shows the data from the human-generated elements. The coefficients for the three mind-sets, the two mind-sets, and the entire panel are displayed in the table, correspondingly.

Index of divergent thought for human-generated elements is 66 and it looks good. The optimum value is around 70. Here we have strong segment to segment differentiation, and many (different elements) high in one segment and low in others.

The outcomes for the AI-generated elements are displayed in Table 9.12.

AI-generated elements for environment. The IDT is 68, almost perfect. We notice a strong segment to segment differentiation; high values in one segment and low in others.

9.7 AI Summarization of the Mind-Sets

The AI-based system, Idea Coach, seamlessly adapts to different scenarios within Bimileap (https://www.bimileap.com/home). It receives two messages: "Create a label for this segment: MS1 AI" (the segment of R54X that has elements of coefficient 21 and higher) and "Describe this segment." In this study, where mind-sets with coefficients of 21 or higher are not present, the system flexibly adjusts and skips the summarization process.

Table 9.11 Data generated from the human-based study

Human-based study	**Total**	**Mindset 1 of 2**	**Mindset 2 of 2**	**Mindset 1 of 3**	**Mindset 2 of 3**	**Mindset 3 of 3**
Base (number of respondents in this group)	**100**	**53**	**47**	**25**	**35**	**40**
Government should consider climate change mitigation and adaptation as one of the most important issues.	12	20	2	19		22
Governments should give more importance to other issues such as industrialization, job creation, and poverty reduction.	14	24	2	16	4	21
I lack sufficient information to provide a conclusive response about climate change mitigation and adaptation.	13	19	5	10	4	21
My life is full of activity, and I don't have time to think about climate change.	15	22	9	19	6	20
Biodiversity conservation should be considered as a significant component of environmental policies.	19	18	19	0	21	26
Biodiversity conservation is a long-term issue, so the focus must be on human well-being, resource	16	11	23		25	19

Table 9.12 Data generated from the AI-based study

	Total	**Mindset 1 of 2**	**Mindset 2 of 2**	**Mindset 1 of 3**	**Mindset 2 of 3**	**Mindset 3 of 3**
Base (number of respondents in this group)	**100**	**55**	**45**	**48**	**28**	**24**
Climate change mitigation and adaptation are the most pressing issues as it poses a significant threat to ecosystems, human health, and the economy.	14	19	7	18	14	3
While climate change is important, we must balance environmental concerns with economic realities.	13	17	10	17	17	2
For many countries, immediate issues such as poverty, access to clean water, and basic healthcare are more pressing than climate change mitigation.	12	19	2	21	12	
The extent of human impact on climate change is overstated, and that	15	21	6	22	16	

9.8 Conclusions

This chapter compares two experiments in the area of governmental policies using the same set of pillars. Researchers and AI create separate elements/answers for each pillar. The goal is to study whether the fact that AI is used to formulate the elements for each pillar in one of the experiments will make a difference in the final results obtained.

The Index of Divergent Thought (IDT) is used (Moskowitz et al. 2020) to evaluate both studies' performances. Experience shows that IDT values around 70 or so reflect stronger segmentations and more dissimilar clustering. In this study, the value of IDT for the AI-based experiment is 68, and for the human-based experiment, it is 66. The AI- and human-generated elements may perform very similarly. These results show that AI-based and human-based experiments can be used interchangeably.

References

Abubakar, Ismaila Rimi et al. 2022. "Environmental Sustainability Impacts of Solid Waste Management Practices in the Global South." *International Journal of Environmental Research and Public Health* 19(19): 12717. https://www.mdpi.com/1660-4601/19/19/12717.

Alestra, Claire, Gilbert Cette, Valérie Chouard, and Rémy Lecat. 2023. "How Can Technology Significantly Contribute to Climate Change Mitigation?" *Applied Economics*: 1–13. https://www.tandfonline.com/doi/full/10.1080/00036846.2023.2227416.

Awewomom, Jonathan et al. 2024. "Addressing Global Environmental Pollution Using Environmental Control Techniques: A Focus on Environmental Policy and Preventive Environmental Management." *Discover Environment* 2(1): 8. https://link.springer.com/10.1007/s44274-024-00033-5.

Clémençon, Raymond. 2016. "The Two Sides of the Paris Climate Agreement." *The Journal of Environment & Development* 25(1): 3–24. http://journals.sagepub.com/doi/10.1177/1070496516631362.

Gofman, A., and H. Moskowitz. 2012. *Rule Developing Experimentation: A Systematic Approach to Understand and Engineer the Consumer Mind.* eds. Alex Gofman (Deceased) and Howard R. Moskowitz. BENTHAM SCIENCE PUBLISHERS.

Gofman, A., and H.R. Moskowitz. 2010. "Isomorphic Permuted Experimental Designs And Their Application In Conjoint Analysis." *Journal of Sensory Studies* 25(1): 127–45.

Hafez, Fatma S. et al. 2023. "Energy Efficiency in Sustainable Buildings: A Systematic Review with Taxonomy, Challenges, Motivations, Methodological Aspects, Recommendations, and Pathways for Future Research." *Energy Strategy Reviews* 45: 101013. https://linkinghub.elsevier.com/retrieve/pii/S2211467X22002073.

Hajam, Younis Ahmad, Rajesh Kumar, and Ajay Kumar. 2023. "Environmental Waste Management Strategies and Vermi Transformation for Sustainable Development." *Environmental Challenges* 13: 100747. https://linkinghub.elsevier.com/retrieve/pii/S2667010023000719.

Huang, Meng-Tian, and Pan-Mao Zhai. 2021. "Achieving Paris Agreement Temperature Goals Requires Carbon Neutrality by Middle Century with Far-Reaching Transitions in the Whole Society." *Advances in Climate Change Research* 12(2): 281–86. https://linkinghub.elsevier.com/retrieve/pii/S1674927821000435.

Ilollari, O., Papajorgji, P., & Civici, A. (2020). Tourism in Albania and its challenges during and after Covid-19. Proceedings of the 3rd UNICART Interdisciplinary International Conference.

Jurkus, Egidijus, Ramūnas Povilanskas, and Julius Taminskas. 2022. "Current Trends and Issues in Research on Biodiversity Conservation and Tourism Sustainability." *Sustainability* 14(6): 3342. https://www.mdpi.com/2071-1050/14/6/3342.

Kilpatrick, A. Marm, Daniel J. Salkeld, Georgia Titcomb, and Micah B. Hahn. 2017. "Conservation of Biodiversity as a Strategy for Improving Human Health and Well-Being." *Philosophical Transactions of the Royal Society B: Biological Sciences* 372(1722): 20160131. https://royalsocietypublishing.org/doi/10.1098/rstb.2016.0131.

Kim, Yoomi. 2021. "Technological Innovation, the Kyoto Protocol, and Open Innovation." *Journal of Open Innovation: Technology, Market, and Complexity* 7(3): 198. https://linkinghub.elsevier.com/retrieve/pii/S2199853122009672.

Kumar, Ashutosh et al. 2023. "A Critical Review on Sustainable Hazardous Waste Management Strategies: A Step towards a Circular Economy." *Environmental Science and Pollution Research* 30(48): 105030–55. https://link.springer.com/10.1007/s11356-023-29511-8.

Manne, Alan S., and Richard G. Richels. 2000. "The Kyoto Protocol: A Cost-Effective Strategy for Meeting Environmental Objectives?" In *Efficiency and Equity of Climate Change Policy*, Dordrecht: Springer Netherlands, 43–61. http://link.springer.com/10.1007/978-94-015-9484-4_3.

Milutinovic, V., and J. Salom. 2016. *A Guide to Data-Driven Marketing Strategy*. 1st ed. Ed. SpringerBriefs in Business.

Moskowitz, Howard R., Alex Gofman, Jacqueline Beckley, and Hollis Ashman. (2006). "Founding a New Science: Mind Genomics." *Journal of Sensory Studies*, 21(3), 266–307. https://onlinelibrary.wiley.com/doi/10.1111/j.1745-459X.2004.00066.x.

Moskowitz, H. R., Wren, J., & Papajorgji, P. (2020). Mind genomics and the law. LAP LAMBERT Academic Publishing.

Mucherino, A., P. Papajorgji, and P. Pardalos. 2009. "Clustering by K-Means." In *Data Mining in Agriculture*, 47–82.

Mulatu, Kalkidan A. et al. 2024. "Nationally Determined Contributions to the 2015 Paris Agreement Goals: Transparency in Communications from Developing Country Parties." *Climate Policy* 24(2): 211–27. https://www.tandfonline.com/doi/full/10.1080/14693062.2023.2285519.

OpenAI. (2023). ChatGPT [Large language model]. OpenAI. https://openai.com/chatgpt

Paris Agreement. 2024. https://www.un.org/en/climatechange/paris-agreement (September 4, 2024).

Pimm, Stuart L. 2021. "What Is Biodiversity Conservation?" *Ambio* 50(5): 976–80. https://link.springer.com/10.1007/s13280-020-01399-5.

Raiser, Kilian, Ulrike Kornek, Christian Flachsland, and William F Lamb. 2020. "Is the Paris Agreement Effective? A Systematic Map of the Evidence." *Environmental Research Letters* 15(8): 083006. https://iopscience.iop.org/article/10.1088/1748-9326/ab865c.

Skjærseth, Jon Birger, Steinar Andresen, Guri Bang, and Gørild M. Heggelund. 2021. "The Paris Agreement and Key Actors' Domestic Climate Policy Mixes: Comparative Patterns." *International Environmental Agreements: Politics, Law and Economics* 21(1): 59–73. http://link.springer.com/10.1007/s10784-021-09531-w.

Todri, A., Papajorgji, P., Moskowitz, H., & Scalera, F. (2021). Perceptions regarding distance learning in higher education, smoothing the transition. *Contemporary Educational Technology, 13*(1), 287. https://doi.org/10.30935/cedtech/9274

Ukaogo, Prince O., Ugochukwu Ewuzie, and Chibuzo V. Onwuka. 2020. "Environmental Pollution: Causes, Effects, and the Remedies." In *Microorganisms for Sustainable Environment and Health*, Elsevier, 419–29. https://linkinghub.elsevier.com/retrieve/pii/B9780128190012000218.

United Nations. 1997. *KYOTO PROTOCOL TO THE UNITED NATIONS FRAMEWORK CONVENTION ON CLIMATE CHANGE*. https://unfccc.int/kyoto_protocol.

Vedanarayanan, V. et al. 2022. "Utilization of Sustainable Resources for Promoting Energy Efficiency in Environment Using Smart Technologies" ed. V. Mohanavel. *International Journal of Photoenergy* 2022: 1–9. https://www.hindawi.com/journals/ijp/2022/6711300/.

VICTOR, DAVID G. 2011. *The Collapse of the Kyoto Protocol and the Struggle to Slow Global Warming*. Princeton University Press. http://www.jstor.org/stable/10.2307/j.ctt7t8pq.

Wang, Yihan, Shaojie Zhang, and Shilin Xu. 2022. "Impact of Efficient Resource Management Practices on Sustainable Performance: Moderating Role of Innovative Culture-Evidence From Oil and Gas Firms." *Frontiers in Psychology* 13. https://www.frontiersin.org/articles/10.3389/fpsyg.2022.938247/full.

Wood, Chelsea L. et al. 2014. "Does Biodiversity Protect Humans against Infectious Disease?" *Ecology* 95(4): 817–32. https://esajournals.onlinelibrary.wiley.com/doi/10.1890/13-1041.1.

———. 2017. "Human Infectious Disease Burdens Decrease with Urbanization but Not with Biodiversity." *Philosophical Transactions of the Royal Society B: Biological Sciences* 372(1722): 20160122. https://royalsocietypublishing.org/doi/10.1098/rstb.2016.0122.

Wu, Shuyao et al. 2022. "Promoting Biodiversity Conservation Requires a Better Understanding of the Relationships Between Ecosystem Services and Multiple Biodiversity Dimensions." *Frontiers in Ecology and Evolution* 10. https://www.frontiersin.org/articles/10.3389/fevo.2022.891627/full.

Zdaniuk, Bozena. 2014. "Ordinary Least-Squares (OLS) Model." In *Encyclopedia of Quality of Life and Well-Being Research*, Dordrecht: Springer Netherlands, 4515–17. http://link.springer.com/10.1007/978-94-007-0753-5_2008.

Zemel, R. et al. 2019. "Consumers, and Dairy: Applying Artificial Intelligence, Mind Genomics, and Predictive Viewpoint Typing." In *Handbook of Modern Dairy Science and Technology*, IntechOpen.

Chapter 10
Experiment in Government Policies

10.1 Introduction

Government regulations aim to protect the efficiency, equity, and safety of different economic sectors. They also seek to uphold financial market stability, encourage competition, and safeguard consumers. Government regulations are not just about safeguarding the interests of workers, consumers, and the general public. They are about fostering innovation and economic growth. They are about striking a delicate balance, ensuring that all interests are considered and protected, thereby creating a harmonious economic ecosystem.

The current literature includes studies that explore the multifaceted meaning, potential impact, and social praxis of regulatory governance. Research has shown that government policies and regulations can promote or hinder innovation (Patanakul and Pinto 2014). Stringent and focused policies and regulations can potentially stimulate significant and fundamental changes in product and process technology. However, policies and regulations can also create obstacles and restrictions that sometimes hinder innovation (Zhang et al. 2023; Graafland and Bovenberg 2020; Christainsen and Haveman 1982).

While societies in America and Europe are far more affluent now than they were a century ago, they are also far more regulated. We now reside in homes and apartment complexes where every aspect of construction is strictly regulated, from zoning to material selection to fire safety. We consume food that has been farmed using highly regulated hormones and fertilizers, processed in factories using technology that are subject to public oversight, and sold in regulated establishments with comprehensive labeling and warnings (Shleifer 2005).

Thus, frameworks for regulatory governance are becoming fundamental components of global civilization. (Hinterleitner et al. (2023) offer a conceptual framework for situating past and future research efforts and facilitating the engagement between them. Comprehensive governance frameworks that regulate and channel a

P. Papajorgji, H. Moskowitz, *The Mind of Everyday*,
https://doi.org/10.1007/978-3-031-78078-3_10

range of social transactions, including economic, political, legal, and cultural ones on a global scale, have evolved from supply chains to the regimes governing international organizations (Kjaer and Vetterlein 2018; (Asadullah et al. 2018).

SMEs (small and medium enterprises) are facing increased technology demands due to the digital revolution. How the government can assist SMMEs in implementing digital transformation (DT) has grown in importance as a research topic because of factors such as low resource investment, high transformation costs, a lack of talent, and unclear transformation pathways. Zhu et al. (2023) suggest a governance structure and a three-party evolutionary game model for successful DT tactics.

Designing and, later, implementing the right policies is a multifaceted issue. The deliberate process of connecting clearly stated policy goals with policy instruments is now widely understood to be the definition of policy design (Howlett 2021). This is a result of our methodical effort to examine how targets respond to or alter their behavior in response to governance tools. Creating policies after learning about the interactions between instruments and targets is the next step in effective design. When these policies are well-designed, they can then consistently result in the intended policy outcomes, giving our work greater certainty (Cairney 2021; Cairney 2023).

The aspirations of vast masses must be implemented for the designed policies to be representative of all population strata in a nation (Persson 2021). When assessing the ability of political and governmental actors to create and carry out successful plans, it is imperative to take into account the opinions and needs of a significant portion of the populace (Mukherjee et al. 2021).

Being able to gather the views of the majority of the population is essential to designing effective policies. Putting such a concept into practice would require setting up in-depth interviews with individuals from various demographic groups and geographic locations around the nation. These days, dedicated systems exist for gathering data, processing it, and producing comprehensive findings (Popiel 2022).

Studies that address the influence of Green parties on redistribution policies have been conducted. Röth and Schwander (2021) investigate the influence of Green government participation on three aspects of distributive policy making: social consumption, social investment, and taxation policies. They analyze a comprehensive cross-national data collection of 34 countries from 1970 to 2015. It is discovered that when Green parties are part of national governments, more money is spent on social investments, but social consumption and taxation remain unchanged. However, the Greens in government avoid cutting social consumption and lowering corporate and top marginal income taxes since they secure center-left majorities.

A relevant aspect of the government policies that effect considerably people's well-being are the redistributive policies. The redistributive role of government entails the reallocation of resources and wealth across society, particularly focused at eliminating inequities and delivering public goods and services (Costas-Fernández and Lodato 2022). This responsibility is executed through many mechanisms such as taxation, social welfare programs, subsidies and grants, regulation and price controls, public goods provision (Cai and Yue 2020).

By these approaches, the government plays a significant role in ensuring social fairness, economic stability, and equal opportunities for all residents.

Based on the findings presented above, the focus of the experiments on government policies will focus on the following issues:

- Government regulatory policies.
- Government constituent policies.
- Government distributive policies.
- Government redistributive policies.

10.2 The Mind Genomics Experiments

The purpose of this chapter is to examine the outcomes of two different approaches to the same issue—government policies. While the second method will make use of recommendations from an AI-based system, the first will draw on the experience of a human researcher. To determine whether using AI-based expertise instead of human expertise makes a difference, the outcomes will be examined individually. Both research employed mind genomics as their technique (Moskowitz et al. 2006; Milutinovic and Salom 2016; Gofman and Moskowitz 2012).

Any Mind Genomics-based research project begins with an analysis of the problem to determine the most important issues that satisfy the study's objectives (Todri et al. 2020; (Papajorgji et al. 2021). The same government policy issues will be looked at in both studies because the concept of silos is the same in both. Rather, each silo will receive a different set of answers; for instance, in one study, CHATGPT 4 (OpenAI 2024) would provide the answers, whereas in the second study, a human researcher will do so. Each study will be subjected to Mind Genomics analysis, and the results are examined to see whether or not using an AI-powered tool to provide the answers has an impact.

The same technology, https://www.bimileap.com/, was used for both investigations, as detailed in Chap. 6. Both investigations used a 4 × 4 Mind Genomics model in their design. Around 2400 observations were used for each study's evaluation. It was demonstrated that care had a comparable participant distribution to assess the data reliably.

At the outset, participants get the following information outlining the objectives of the study:

> ***Government policies concern everyone, and this study covers issues such as regulatory, constituent, distributive, and redistributive policies. We want your opinion to understand the focus of government policies' issues and challenges.***

The silos and the corresponding answers for each silo for human-based experiment are presented in Table 10.1.

The silos and the corresponding answers for each silo for the AI-based experiment are presented in Table 10.2.

Table 10.1 Elements designed by a human researcher

	Question A: The focus of government regulatory policies
A1	Implement a dynamic regulatory policy framework that can adapt to changing market conditions
A2	Establish a regulatory policy task force composed of industry experts and government officials
A3	Establish a regulatory policy feedback mechanism for businesses and individuals to provide input
A4	Introduce a "regulatory impact assessment" requirement for all proposed policies
	Question B: The focus of government constituent policies
B1	Implement a digital platform for constituents to propose and vote on policy ideas
B2	Create a constituent advisory board to provide feedback on proposed policies
B3	Conduct surveys to gauge constituent priorities and preferences on policy matters
B4	Collaborate with local community to engage constituents in the policy-making process
	Question C: The focus of government distributive policies
C1	Create a tax system that heavily taxes the wealthiest individuals and corporations to redistribute wealth more equitably
C2	Develop a government-funded job guarantee program to ensure full employment and reduce poverty
C3	Introduce a cap on executive salaries to address income inequality and promote fair wages for all employees
C4	Implement a universal basic income program to ensure all citizens have a financial safety net
	Question D: The focus of government redistributive policies
D1	Provide free public transportation to low-income individuals to improve access to jobs and services
D2	Offer free or subsidized childcare to support working families and reduce financial burdens
D3	Establish a progressive tax system that taxes higher incomes at a higher rate to promote economic equity
D4	Create a national affordable housing program to ensure everyone has access to safe and affordable housing

Mind Genomics uses a five-level Likert scale to evaluate vignettes as follows. Level 1 means **Unlikely** and level 5 means **Likely**.

How likely are you to appreciate the focus of government policies if the following is part of it?
1 = not interested ... AND ....won't happen
2 = not interested ... BUT .. It could happen
3 = Can't honestly say
4 = looks interesting .. BUT...won't happen
5 = looks interesting .. AND .. It could happen

Special care was taken to ensure a good distribution of parameters such as age, education level, and gender of participants.

After collecting all the data from the participants, Bimileap will make all calculations and present results in an Excel file containing many sheets as shown in Chap.

Table 10.2 Elements designed by AI-based system

	Question A: The focus of government regulatory policies
A1	Impact of government regulatory policies on small businesses
A2	Effectiveness of government regulatory policies in protecting consumer rights
A3	Transparency and accountability in the implementation of government regulatory policies
A4	Balancing economic growth with environmental protection through government regulatory policies
	Question B: The focus of government constituent policies
B1	Healthcare reform to ensure access and affordability for all citizens
B2	Education policies to improve the quality of public schools and address disparities in resources
B3	Economic development strategies to create jobs and support small businesses
B4	Environmental protection measures to address climate change and promote sustainability
	Question C: The focus of government distributive policies
C1	Distributive policies must include income inequality
C2	Distributive policies must include access to quality education
C3	Distributive policies must include affordable healthcare
C4	Distributive policies must include housing affordability
	Question D: The focus of government redistributive policies
D1	Effectiveness of current redistributive policies
D2	Impact of redistributive policies on income inequality
D3	Public perception and support for redistributive policies

6. Mind Genomics provides a holistic analysis of a problem under study; it allows researchers to create up to ten independent variables and distribute the results over them. One of the parameters Mind Genomics uses is the participant's gender. Although gender is not used in this study, we believe it might be relevant for other studies (Todri et al. 2024).

Following participant data collection, all computations will be performed by Bimileap (https://www.bimileap.com) and the findings will be presented in an Excel file with many sheets, as shown in Chap. 6.

10.3 Comparing Results of Both Experiments: What We Found

The elements produced by AI were different from those produced by researchers, as Fig. 10.1 illustrates. The overall total per silo in the AI-designed experiment in this study had greater values than the human-designed experiment. Comparing the broad patterns that emerged from the two research on the same issue is appropriate, even if we achieved different and higher numbers from the AI-based experiment. The patterns of responses produced by the two experiments can be compared in several ways, including the rating patterns, the relationships between elements' presence or absence and responses to the vignette (modeling), the interpretation of high-performing elements (such as interpretation and AI-based summarization), and, lastly, the way that respondents were clustered into mind-sets and the characteristics of these mind-sets.

After gathering the participant data (see Chap. 6), the Mind Genomics platform performed the necessary computations. Once more, two and three mind-sets were extracted using clustering based on the coefficients of the 16 elements (question answers). The coefficients resulting from the cluster analysis, which only considered the 16 items that participants submitted, are displayed in Tables 10.3 for the human-based experiment and 10.4 for the AI-based experiment. These elements were clustered into two and then three mind-sets, respectively.

Analyzing the data of Table 10.3, the two and three mind-sets for the human-based study, we notice that the highest-scoring value is 21 for the two mind-set solution and 31 for the three mind-set solution. Thus, clusters in the three mind-set solution have higher scoring elements.

Table 10.4 shows the two and three mind-set values for the AI-based experiment.

Analyzing the data of Table 10.4, the two and three mind-sets for the AI-based study, we notice that the highest-scoring value is 27 for the two mind-set solution and 32 for the three mind-set solution. Thus, clusters in the three mind-set solution have higher scoring elements.

Before starting to analyze the data from the two experiments, a conversion from a five-scale evaluation to a binary evaluation is made in order to make the analyses easier, as detailed in Chap. 6. New binary variables are created by the statistical analysis that underlies this. The alteration is displayed in Table 10.5. We are altering the original 5-point scale because it does not provide us with the necessary format to conduct some of the studies that we need to (viz., modeling and clustering).

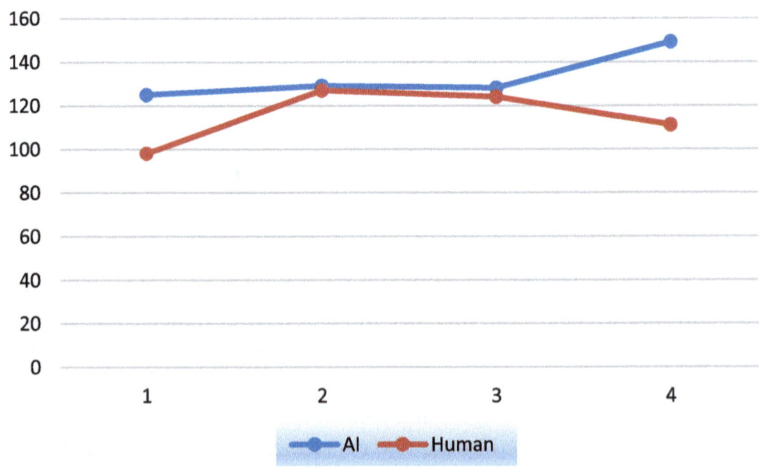

Fig. 10.1 Total silos per experiment

Table 10.3 The two and three mind-set solution for the human-based study

	Group (Binary Ratings) Human generated elements	Total	Mindset 1 of 2	Mindset 2 of 2	Mindset 1 of 3	Mindset 2 of 3	Mindset 3 of 3
	Base Size	100	39	61	25	40	35
	Question A: The focus of government regulatory policies						
A1	Implement a dynamic regulatory policy framework that can adapt to changing market conditions,	14	21	10	25	10	10
A2	Establish a regulatory policy task force composed of industry experts and government officials,	10	16	7	18	8	7
A3	Establish a regulatory policy feedback mechanism for businesses and individuals to provide input,	14	19	11	17	15	9
A4	Introduce a "regulatory impact assessment" requirement for all proposed policies	10	17	6	21	10	1
	Question B: The focus of government constituent policies,						
B1	Implement a digital platform for constituents to propose and vote on policy ideas,	17	14	19	9	21	19
B2	Create a constituent advisory board to provide feedback on proposed policies,	15	16	15	13	18	15
B3	Conduct surveys to gauge constituent priorities and preferences on policy matters	17	14	19	13	18	19
B4	Collaborate with local community to engage constituents in the policy-making process	14	16	13	11	15	15
	Question C: The focus of government distributive policies						
C1	Create a tax system that heavily taxes the wealthiest individuals and corporations to redistribute wealth more equitably	17	17	18	7	31	8
C2	Develop a government-funded job guarantee program to ensure full employment and reduce poverty	14	13	15		28	8
C3	Introduce a cap on executive salaries to address income inequality and promote fair wages for all employees	17	21	15	10	29	9
C4	Implement a universal basic income program to ensure all citizens have a financial safety net.	13	17	12	9	24	5
	Question D: The focus of government redistributive policies.						
D1	Provide free public transportation to low-income individuals to improve access to jobs and services	16	11	18	18	9	22
D2	Offer free or subsidized childcare to support working families and reduce financial burdens,	14	5	20	10	7	28
D3	Establish a progressive tax system that taxes higher incomes at a higher rate to promote economic equity	13	11	15	15	8	19
D4	Create a national affordable housing program to ensure everyone has access to safe and affordable housing	12	7	15	12	7	20

At this study stage, the focus is on the binary variable R54x, generated when ratings 2, 3, and 1 are transformed to 0 while ratings 5 and 4 are converted to 100. A vanishingly small random value (<10–5) is added to this freshly formed binary-dependent variable (BDV). To ensure that R54x has some minimal variability and can be utilized in OLS (ordinary least-squares regression), the value of (<10–5) is inserted as a precaution.

The percentages of rating points allotted to the vignettes receive the most attention. The main focus is on how the rating distributions of elements selected by AI and humans differ from one another. Ratings for responses derived from human-sourced ideas of “5” is higher than ratings for responses derived from other sources (20 for the AI-based study against 25 for the human-based study), as Table 10.6 illustrates. This observation suggests that the two techniques have a little difference in philosophy. This discovery will be significant and have intriguing ramifications, as can be seen here.

Table 10.4 The two and three mind-set solution for the AI-based study

	Group (Binary Ratings Table 2: Elements defined by AI)	**Total**	**Mindset 1 of 2**	**Mindset 2 of 2**	**Mindset 1 of 3**	**Mindset 2 of 3**	**Mindset 3 of 3**
	Base Size	100	46	54	34	38	28
	Question A: The focus of government regulatory policies						
A1	Impact of government regulatory policies on small businesses	16	22	11	24	15	8
A2	Effectiveness of government regulatory policies in protecting consumer rights	16	17	15	16	20	11
A3	Transparency and accountability in the implementation of government regulatory policies	14	12	16	17	23	
A4	Balancing economic growth with environmental protection through government regulatory policies	16	14	18	15	21	12
	Question B: The focus of government constituent policies						
B1	Healthcare reform to ensure access and affordability for all citizens	16	24		21	5	23
B2	Education policies to improve the quality of public schools and address disparities in resources	15	23	8	22	2	23
B3	Economic development strategies to create jobs and support small businesses	15	20	12	17	8	23
B4	Environmental protection measures to address climate change and promote sustainability	17	25	10	23	2	30
	Question C: The focus of government distributive policies						
C1	Distributive policies must include income inequality	16	6	24		21	30
C2	Distributive policies must include access to quality education	18	8	25	6	22	26
C3	Distributive policies must include affordable healthcare	16	2	27		22	32
C4	Distributive policies must include housing affordability	18	6	27		24	30
	Question D: The focus of government redistributive policies						
D1	Effectiveness of current redistributive policies	20	22	18	22	18	17
D2	Impact of redistributive policies on income inequality	21	25	18	22	16	28
D3	Public perception and support for redistributive policies	19	21	17	22	18	13
D4	Potential reforms to improve redistributive policies	21	19	23	16	22	24

In the following step of our analysis, we examine the elements' performance. The significant variable R54x is used to quantify the performance of the two approaches (AI and human based). Scale points 4 and 5 are united to form R54x, which has the attribute "it looks interesting." Ordinary least-squares regression, or OLS for short, is the method that studies the relationship between the elements (Zdaniuk 2014).

For each respondent, the underlying experimental design ensures that the 16 elements are presented in a way that makes them statistically independent of each other and has the property that the 24 vignettes each respondent evaluates are different from the 24 vignettes assessed by the other respondent. The study's setup allows the researcher to create regression equations.

The analysis for each study's total panel creates this simple equation: R54x = k1A1 + k2A2…k16A16.

Table 10.5 Transformation of a 5-rating to a two-rating scale

	The original points on the rating scale				
	1	2	3	4	5
Interested	N	N	DK	Y	Y
Happen	N	Y	DK	N	Y
New binary variables					
R1x	100	0	0	0	0
R2x	0	100	0	0	0
R3x	0	0	100	0	0
R4x	0	0	0	100	0
R5x	0	0	0	0	100
R54x	0	0	0	100	100
R52x	0	100	0	0	100
R41x	100	0	0	100	0
R21x	100	100	0	0	0

Table 10.6 Frequency of scale point per vignette

	Original scale points. Numbers in body of table are percent of times the rating scale was assigned to a vignette					
	1	2	3	4	5	Total
Source						
Total	13	11	22	26	27	100
AI	11	8	22	29	20	100
Human	14	14	22	24	25	100

The coefficients reveal the driving power of each element. According to statistical assessments, these coefficients, k1–k16, achieve statistical significance when they approach or surpass 10–12. The focus is on items with relevance coefficients of 16 or higher and importance coefficients of 21 or higher. Researchers' experience plays a crucial role in determining the coefficients' cutoff level, which is more stringent than a straightforward analysis of the coefficient's T value as determined by the regression model's analyses of variance.

Table 10.7 shows the coefficients of the 16 elements from OLS regression, showing the results from human-generated elements.

Analyzing the data of Table 10.7, we notice no strong-performing elements of 21 and higher. The highest value is 17 (several items have a value of 17), and the lowest value is 10 (again, several items have a value of 10). We notice a small range of coefficients (from 10 to 17).

The data in Table 10.8 show several items with value 21 for element D2: **Impact of redistributive policies on income inequality** and element D4: **Potential reforms to improve redistributive policies**. The minimum value is 14. Even in this case, we notice the small range of values (14 to 21).

Table 10.7 Performance of element R54x for the human-based experiment

	Performance of HUMAN-generated elements (dependent variable = R54x—INTERESTED)	Total
	Base size	100
B1	Implement a digital platform for constituents to propose and vote on policy ideas	17
B3	Conduct surveys to gauge constituent priorities and preferences on policy matters	17
C1	Create a tax system that heavily taxes the wealthiest individuals and corporations to redistribute wealth more equitably	17
C3	Introduce a cap on executive salaries to address income inequality and promote fair wages for all employees	17
D1	Provide free public transportation to low-income individuals to improve access to jobs and services	16
B2	Create a constituent advisory board to provide feedback on proposed policies	15
A1	Implement a dynamic regulatory policy framework that can adapt to changing market conditions	14
A3	Establish a regulatory policy feedback mechanism for businesses and individuals to provide input	14
B4	Collaborate with local community to engage constituents in the policy-making process	14
C2	Develop a government-funded job guarantee program to ensure full employment and reduce poverty	14
D2	Offer free or subsidized childcare to support working families and reduce financial burdens	14
C4	Implement a universal basic income program to ensure all citizens have a financial safety net	13
D3	Establish a progressive tax system that taxes higher incomes at a higher rate to promote economic equity	13
D4	Create a national affordable housing program to ensure everyone has access to safe and affordable housing	12
A2	Establish a regulatory policy task force composed of industry experts and government officials	10
A4	Introduce a "regulatory impact assessment" requirement for all proposed policies	10

10.4 AI Summarization of Strong Elements

An AI-based summarizer utilizing Chap GPT4 (OpenAI 2024) was employed as a component of the Mind Genomics platform to investigate various facets of high-performing parts with a score of 21 or above.

The human-based study results do not include items with a value of 21 or higher, so the AI summarization process does not apply.

In the case of the AI-based experiment, only two elements passed the cut for the AI-generated elements: D2—**Impact of redistributive policies on income inequality** and D4—**Potential reforms to improve redistributive policies** as shown in Table 10.8. Despite only two elements, the AI embedded in Idea Coach returned with a complete analysis of these elements based on a series of predefined queries. The results of these queries are shown in Table 10.9. It may not be necessary to

Table 10.8 shows the performance of element R54x for the AI-generated experiment

	Performance of AI-generated elements (Dependent variable = R54x – INTERESTED)	Total
	Base Size	100
D2	Impact of redistributive policies on income inequality	21
D4	Potential reforms to improve redistributive policies	21
D1	Effectiveness of current redistributive policies	20
D3	Public perception and support for redistributive policies	19
C2	Distributive policies must include access to quality education	18
C4	Distributive policies must include housing affordability	18
B4	Environmental protection measures to address climate change and promote sustainability	17
A1	Impact of government regulatory policies on small businesses	16
A2	Effectiveness of government regulatory policies in protecting consumer rights	16
A4	Balancing economic growth with environmental protection through government regulatory policies	16
B1	Healthcare reform to ensure access and affordability for all citizens	16
C1	Distributive policies must include income inequality	16
C3	Distributive policies must include affordable healthcare	16
B2	Education policies to improve the quality of public schools and address disparities in resources	15
B3	Economic development strategies to create jobs and support small businesses	15
A3	Transparency and accountability in the implementation of government regulatory policies	14

repeat the table's contents, but it is very instructive to see how AI in the form of Chat GPT4 can take a single theme and create an entire "thought piece" around that theme. In a sense, Table 10.9 shows the potential richness that may emerge when the researcher explores the implications of the theme. To Idea Coach, an AI-based system embedded into Bimileap (https://www.bimileap.com/home) sends two messages: "Create a label for this segment" (the segment with the element of coefficient 21) and "Describe this segment." The results of the two messages are shown in Table 10.9.

10.5 Beyond the Single Dependent Variable to Considering the Pattern for All Variables

The Mind Genomics method allows the researcher to evaluate the relative contributions of the various parts to a range of binary-dependent variables. Since we have only examined one dependent variable thus far—R54x—we know how closely different elements perform in terms of R54x, except AI-generated items, which perform marginally better than human-generated ones.

The meticulous modeling and analysis process, which remained consistent throughout the study, allowed us to examine the spectrum of coefficients for each of the binary-dependent variables displayed in Table 10.5.

To gain a deeper understanding of the differences between AI-generated and human-generated elements, we use OLS regression to establish a relationship between the newly produced BDVs and response time and the presence or absence of elements. Table 10.10 shows the coefficients for the AI-generated elements.

Table 10.9 Information received from Idea Coach; The AI-based system embedded into Bimileap

AI-generated elements—Summarizer for the Total Panel **Overall—Top** The following is a description of attributes from the total sample of a research study. Interested in: Impact of redistributive policies on income inequality Potential reforms to improve redistributive policies
Create a label for this segment: "Redistributive policy impact and reform recommendations"
Describe this segment: The total sample of the research study consists of individuals from various socio-economic backgrounds and demographic profiles. They are likely to have diverse views on income inequality and redistributive policies. Some may be directly impacted by these policies, either benefiting from or feeling burdened by them. Others may have a general interest in economic equality and social justice. This segment of the sample is likely to be open to potential reforms that aim to improve redistributive policies and address income inequality. They may have varying levels of knowledge and understanding regarding economic systems and policies but are willing to engage in discussions and debates on these topics. Overall, this segment represents a diverse and engaged group of individuals interested in the impact and potential improvements of redistributive policies
Describe the attractiveness of this segment as a target audience: This segment of the total sample is highly attractive as a target audience for several reasons. Firstly, they are interested in a topic that is highly relevant and important in today's society—Income inequality and redistributive policies. This indicates that they are likely to be engaged and receptive to information and messaging related to this issue
Additionally, their interest in potential reforms to improve redistributive policies suggests that they are proactive and interested in finding solutions to address income inequality. This makes them an ideal audience for organizations or individuals looking to advocate for policy change or mobilize support for specific reform initiatives
Overall, this segment's combination of interest in the topic, willingness to engage in dialogue around potential solutions, and likely alignment with progressive or reformist views on redistributive policies makes them a highly desirable target audience for anyone looking to drive change in this area

(continued)

Table 10.9 (continued)

Explain why this segment might not be attractive as a target audience:
This segment may not be attractive as a target audience because they may already have strong opinions or beliefs about redistributive policies and potential reforms. They may be less receptive to new information or perspectives, making it challenging to engage in meaningful dialogue or discussion. Additionally, this segment may be limited in terms of their willingness to consider alternative viewpoints or be open to changing their views, hindering the ability to effectively communicate research findings or policy recommendations. Targeting this segment may therefore be less effective in terms of promoting dialogue, generating support for policy reforms, or influencing attitudes and behavior

List what is missing or should be known about this segment, in question form:
– What specific redistributive policies were implemented in the study?
– How were income inequality levels measured before and after the implementation of redistributive policies?
– Were there any control variables taken into account when analyzing the impact of redistributive policies on income inequality?
– Did the study consider any potential unintended consequences of the redistributive policies on other aspects of the economy?
– Were there any limitations or biases in the data used to analyze the effects of redistributive policies on income inequality?
– What were the demographics of the sample population used in the study?
– How long were the redistributive policies in place before their impact on income inequality was assessed?
– Did the study assess the effectiveness of different types of redistributive policies in reducing income inequality?
– Were there any differences in the impact of redistributive policies on income inequality based on income level or socioeconomic status?
– Were there any comparisons made between the effectiveness of redistributive policies in different countries or regions?

List and briefly describe attractive new or innovative products, services, experiences, or policies for this segment:

1. Universal basic income: Implementing a universal basic income policy that provides all citizens with a minimum income could help reduce income inequality and ensure financial security for all individuals
2. Wealth tax: Introducing a wealth tax on high-net-worth individuals could help redistribute wealth and reduce income inequality in society
3. Guaranteed minimum income: Establishing a guaranteed minimum income for all individuals, regardless of their employment status, could help alleviate poverty and reduce income inequality
4. Affordable housing initiatives: Implementing affordable housing initiatives such as rent control or subsidized housing could help lower-income individuals afford housing and reduce inequality in access to housing
5. Access to education and healthcare: Providing free or affordable access to education and healthcare for all individuals could help improve social mobility and reduce income inequality in the long run

(continued)

Table 10.9 (continued)

Which messages will interest this segment? 1. Concerned about income inequality: This segment of the population is likely to be interested in messages that address the issue of income inequality and the impact of redistributive policies on closing the wealth gap 2. Open to potential reforms: This group may be receptive to messages that discuss potential reforms to improve redistributive policies, such as implementing more progressive tax policies or increasing funding for social welfare programs 3. Interested in social justice: Messages that emphasize the importance of social justice and fairness in income distribution are likely to resonate with this segment 4. Looking for solutions: This group may be interested in messages that offer concrete solutions to address income inequality, such as advocating for a living wage or expanding access to affordable housing 5. Interested in economic policy: Messages that provide insights into the economic implications of redistributive policies and potential reforms may capture the attention of this segment Overall, messages that emphasize the importance of addressing income inequality, propose actionable solutions, and highlight the potential impact of redistributive policies are likely to interest this segment

Table 10.11 shows the coefficients for the elements that were created by humans. Tables show darkened cells that draw attention to the underlying patterns. The shading criteria that are operationally specified are as follows:

R54x—Shaded when the coefficient is 21 or higher.
Remaining BDV—Shaded when the coefficient is 15 or higher.
RT—Shaded when the response time is 1.2 or higher.

Table 10.10 shows the coefficients for binary-dependent variables and RT, for AI-generated elements.

Table 10.10 shows that we have only two elements with values 21 or higher corresponding to dependent variable R54x and several values of 1.2 and higher corresponding to RT.

Table 10.11 shows that there are no values equal or higher than 21 for the dependent variables and several values equal or higher than 1.2 for RT.

10.6 Moving from the Total Panel to the Mind-Sets, and to the IDT (Index of Divergent Thought)

Mind genomics distinguishes itself by focusing on mind-sets developed from the granularity of the data and on specific challenges rather than general subjects. Mind Genomics investigates the pattern of responses to particular themes, such as those discussed here, starting from the bottom up.

Using this little "slice" of reality, Mind Genomics aims to identify interpretable and potentially meaningful groups of people displaying different responses to the

Table 10.10 Coefficients for binary-dependent variables and RT, for AI-generated elements

		Newly created binary dependent variables and Response Time (RT)									
Factor	**AI generated elements**	R54X	R52X	R41X	R21X	R1X	R2X	R3X	R4X	R5X	RTX
A1	Impact of government regulatory policies on small businesses	16	8	8	4	3	1	8	8	7	1.1
A2	Effectiveness of government regulatory policies in protecting consumer rights	16	9	8	4	3	1	9	8	8	1.0
A3	Transparency and accountability in the implementation of government regulatory policies	14	9	7	5	3	3	7	7	6	1.1
A4	Balancing economic growth with environmental protection through government regulatory policies	16	9	8	5	4	1	8	8	8	1.0
B1	Healthcare reform to ensure access and affordability for all citizens	15	13	6	7	2	4	8	6	9	1.3
B2	Education policies to improve the quality of public schools and address disparities in resources	15	9	8	5	4	2	9	8	7	1.4
B3	Economic development strategies to create jobs and support small businesses	15	12	7	7	3	4	8	7	8	1.4
B4	Environmental protection measures to address climate change and promote sustainability	17	8	10	6	5	2	8	10	7	1.5
C1	Distributive policies must include income inequality	16	11	8	6	3	3	8	8	8	1.0
C2	Distributive policies must include access to quality education	18	14	6	7	4	3	6	6	12	1.2
C3	Distributive policies must include affordable healthcare	16	9	10	6	4	2	7	10	7	1.0
C4	Distributive policies must include housing affordability	18	10	11	7	4	3	5	11	7	1.1
D1	Effectiveness of current redistributive policies	20	13	8	2	1	1	7	8	12	1.3
D2	Impact of redistributive policies on income inequality	21	13	12	7	3	4	1	12	9	1.2
D3	Public perception and support for redistributive policies	18	15	7	7	3	3	3	7	12	1.2
D4	Potential reforms to improve redistributive policies	21	14	10	5	2	3	3	10	11	1.2

same stimuli. These ideas originate from the actual investigation. Their answers are all different.

The patterns are interpretable and suggest different people's "mental realities." Furthermore, only statistical methods are employed to identify these mind-sets. Finally, there are ways to categorize recently hired people who have yet to participate in the study into the respective mind-sets that have been discovered. As a result, these approaches use several questions from the first study and enable the growth of mind-sets (Zemel et al. 2019).

The method for discovering, or perhaps better, creating these mind-sets uses k-means clustering (Mucherino et al. 2009). The underlying experimental design permits the creation of 24 vignettes in precisely the mathematical structure needed for OLS (ordinary least-squares) regression. We saw the use of OLS above for the data from the total panels, both for people responding to AI-generated elements and separately for people responding to human-generated elements. We did this for the total panel of 100 respondents, putting all the data together to create a dataset of 2400 rows.

Following the initial clustering, a more detailed analysis is conducted using OLS regression. Each respondent's 24 vignettes undergo the same OLS regression, a crucial step in our analysis process. The k-means clustering methodology is then used to calculate the "distance" between each pair of respondents. The amount (1-Pearson R) separating the two matched sets of 16 coefficients is the definition of "distance" for the majority of Mind Genomics investigations.

Table 10.11 Coefficients for binary-dependent variables and RT, for human-generated elements

		Newly created binary dependent variables and Response Time (RT)									
Factor	**Human generated elements**	R54X	R52X	R41X	R21X	R1X	R2X	R3X	R4X	R5X	RTX
A1	Implement a dynamic regulatory policy framework that can adapt to changing market conditions,	14	12	6	8	5	4	7	6	8	1.4
A2	Establish a regulatory policy task force composed of industry experts and government officials,	10	11	6	10	3	7	7	6	4	1.1
A3	Establish a regulatory policy feedback mechanism for businesses and individuals to provide input,	14	8	8	8	6	3	6	8	6	1.3
A4	Introduce a "regulatory impact assessment" requirement for all proposed policies	10	11	4	9	3	5	10	4	6	1.2
B1	Implement a digital platform for constituents to propose and vote on policy ideas,	17	12	7	6	4	2	7	7	10	1.4
B2	Create a constituent advisory board to provide feedback on proposed policies,	15	14	5	6	3	3	8	5	11	1.1
B3	Conduct surveys to gauge constituent priorities and preferences on policy matters	17	12	6	5	5	1	8	6	11	1.4
B4	Collaborate with local community to engage constituents in the policy-making process	14	14	5	8	3	6	8	5	9	1.5
C1	Create a tax system that heavily taxes the wealthiest individuals and corporations to redistribute wealth more equitably	17	13	8	9	5	4	5	8	9	1.3
C2	Develop a government-funded job guarantee program to ensure full employment and reduce poverty	14	12	7	11	6	5	5	7	7	1.2
C3	Introduce a cap on executive salaries to address income inequality and promote fair wages for all employees	17	12	9	11	7	4	3	9	9	1.5
C4	Implement a universal basic income program to ensure all citizens have a financial safety net.	13	13	9	11	3	8	6	9	5	1.4
D1	Provide free public transportation to low-income individuals to improve access to jobs and services	16	12	8	7	2	5	5	8	7	1.4
D2	Offer free or subsidized childcare to support working families and reduce financial burden s,	14	5	10	7	6	0	5	10	4	1.3
D3	Establish a progressive tax system that taxes higher incomes at a higher rate to promote economic equity	13	13	5	7	2	4	6	5	8	1.1
D4	Create a national affordable housing program to ensure everyone has access to safe and affordable housing	12	7	8	9	5	3	7	8	4	1.2

If both responders exhibit similar patterns, the distance equals 0 (1–1 = 0), and the Pearson R is 1. The distance increases to 2 (1- – 1 = 2) when the two respondents exhibit inverse patterns and move in opposite directions. The Pearson R value is −1 in this scenario. The clustering approach divides the responses into two and then three groups, or clusters, after determining the distances. These clusters are referred to as mind-sets in the terminology of mind genomics. The thoughts are distinct from one another and uniform. Pairs of responders had tiny "distances" inside a mentality or cluster.

The data from the AI-generated elements are shown in Table 10.12. It offers the coefficients for the three mind-sets, the two mind-sets, and the entire panel, in that order. Comparably, the outcomes for the human-generated elements are shown in Table 10.13. A thorough comprehension of the outcomes is made possible by the data's lucid presentation.

The three mind-set solution is ranked first since it will be the one chosen to display the various emergent groups. The items are arranged in each table in descending order. In each group, just the shaded elements for mind-sets 1 (MS1), 2 and 3 are displayed. Lastly, a piece that works well for two perspectives will display twice, once in the appropriate spot for each mind-set.

While analyzing Tables 10.12 and 10.13, we notice that the elements created by AI generate higher coefficients than those created by humans. Our first conclusion, therefore, is that the AI elements generate stronger responses. However, a deeper pattern in the data may cast a shadow on the clearly stronger performance of the elements across mind-sets. This is the conclusion reached after computing the IDT, the index of divergent thought (Thakral et al. 2021; Madore et al. 2016).

The IDT is used to gauge how successfully the Mind Genomics initiative gathered information about people's thoughts. Initially, the idea was that high coefficients were preferable to low coefficients. The IDT was proposed as a computational formula that demonstrated how well the ideas worked. The computational formula appears at the bottom of Table 10.12 for AI-generated items, while for human-generated elements, it appears at the bottom of Table 10.13. The IDT measures the elements' performance (Giancola et al. 2023). The IDT's initial readings were frequently in the range of 50–70. Since the elements would have greater coefficients, it was thought that the learning would be demonstrated with higher IDT values.

However, it was also becoming apparent that very high IDT values corresponded to elements that performed well across all mind-sets. Simulations suggested that IDT values around 70 corresponded to strong-performing elements in some mind-sets and not in others, a finding that is more meaningful than having strong-performing elements in all mind-sets.

Table 10.12 AI-generated IDT values

AI generated IDT	**Total**	**Mindset 1 of 2**	**Mindset 2 of 2**	**Mindset 1 of 3**	**Mindset 2 of 3**	**Mindset 3 of 3**
Base (number of respondents in this group)	**100**	**46**	**54**	**34**	**38**	**28**
Impact of government regulatory policies on small businesses	16	22	11	24	15	8
Effectiveness of government regulatory policies in protecting consumer rights	16	17	15	16	20	11
Transparency and accountability in the implementation of government regulatory policies	14	12	16	17	23	
Balancing economic growth with environmental protection through government regulatory policies	16	14	18	15	21	12
Healthcare reform to ensure access and affordability for all citizens	16	24	8	21	5	23
Education policies to improve the quality of public schools and address disparities in resources	15	23	8	22	2	23
Economic development strategies to create jobs and support small businesses	15	20	12	17	8	23
Environmental protection measures to address climate change and promote sustainability	17	25	10	23	2	30
Distributive policies must include income inequality	16	6	24	0	21	30
Distributive policies must include access to quality education	18	8	25	6	22	26
Distributive policies must include affordable healthcare	16	2	27		22	32
Distributive policies must include housing affordability	18	6	27	0	24	30
Effectiveness of current redistributive policies	20	22	18	22	18	17
Impact of redistributive policies on income inequality	21	25	18	22	16	28
Public perception and support for redistributive policies	19	21	17	22	18	13
Potential reforms to improve redistributive policies	21	19	23	16	22	24
Sum Total of Squares of All Positive Coefficients	**4762**	**5294**	**5423**	**4833**	**5065**	**8134**
Average of All Positive Coefficients (sum coefficients/base)	**47.6**	**115.1**	**100.4**	**142.1**	**133.3**	**290.5**
Number of respondents in the column (in the subgroup)	**100**	**46**	**54**	**34**	**38**	**28**
Weight = Proportion of the respondents in the subgroup	**0.33**	**0.15**	**0.18**	**0.11**	**0.13**	**0.09**
Weighted Total (Weight x Sum Total of Squares of All Positive Coefficients)	**1571.5**	**794.1**	**976.14**	**531.63**	**658.45**	**732.06**
Final Score = sum of weighted total	**5263.8**	**73**				

Table 10.13 Human-generated IDT value

	Total	Mindset 1 of 2	Mindset 2 of 2	Mindset 1 of 3	Mindset 2 of 3	Mindset 3 of 3
Base (number of respondents in this group)	**100**	**39**	**61**	**25**	**40**	**35**
Implement a dynamic regulatory policy framework that can adapt to changing market conditions,	14	21	10	25	10	10
Establish a regulatory policy task force composed of industry experts and government officials,	10	16	7	18	8	7
Establish a regulatory policy feedback mechanism for businesses and individuals to provide input,	14	19	11	17	15	9
Introduce a "regulatory impact assessment" requirement for all proposed policies	10	17	6	21	10	1
Implement a digital platform for constituents to propose and vote on policy ideas,	17	14	19	9	21	19
Create a constituent advisory board to provide feedback on proposed policies,	15	16	15	13	18	15
Conduct surveys to gauge constituent priorities and preferences on policy matters	17	14	19	13	18	19
Collaborate with local community to engage constituents in the policy-making process	14	16	13	11	15	15
Create a tax system that heavily taxes the wealthiest individuals and corporations to redistribute wealth more equitably	17	17	18	7	31	8
Develop a government-funded job guarantee program to ensure full employment and reduce poverty	14	13	15		28	8
Introduce a cap on executive salaries to address income inequality and promote fair wages for all employees	17	21	15	10	29	9
Implement a universal basic income program to ensure all citizens have a financial safety net.	13	17	12	9	24	5
Provide free public transportation to low-income individuals to improve access to jobs and services	16	11	18	18	9	22
Offer free or subsidized childcare to support working families and reduce financial burdens,	14	5	20	10	7	28
Establish a progressive tax system that taxes higher incomes at a higher rate to promote economic equity	13	11	15	15	8	19
Create a national affordable housing program to ensure everyone has access to safe and affordable housing	12	7	15	12	7	20
Sum Total of Squares of All Positive Coefficients	**3299**	**3755**	**3514**	**3242**	**5208**	**3666**
Average of All Positive Coefficients (sum coefficients/base)	**33**	**96.3**	**57.6**	**129.7**	**130.2**	**104.7**
Number of respondents in the column (in the subgroup)	**100**	**39**	**61**	**25**	**40**	**35**
Weight = Proportion of the respondents in the subgroup	**0.33**	**0.13**	**0.2**	**0.08**	**0.13**	**0.12**
Weighted Total (Weight x Sum Total of Squares of All Positive Coefficients)	**1088.7**	**488.15**	**702.8**	**259.36**	**677.04**	**439.92**
Final Score = sum of weighted total	**3655.9**	**60**				

Based on the newly developed rule of thumb that IDT values of 70 or so reflect stronger segmentations and more dissimilar clustering, we are left with the interesting result that the IDT value for AI-generated elements is 73. In contrast, the IDT value for human-suggest elements is 60.

Table 10.12 shows the coefficients of AI-generated elements and IDT.

The IDT value for the AI-generated elements is 73. This value shows good differentiation across mind-sets. Table 10.13 shows the coefficients of AI-generated elements and IDT.

The IDT value for the human-based experiment is 60; this value is not optimal, and it shows that there is less mind-set differentiation.

10.7 AI Summarization of the Mind-Sets

The following presents the dialog with the AI system embedded into Idea Coach (https://www.bimileap.com/), describing the queries and answers for the human-based experiment.

The AI system was required to create a label for the segment containing elements Of 54x coefficient with values 21+. There are only two elements with values 21 or higher: **Implement a dynamic regulatory policy framework that can adapt to changing market conditions and Introduce a cap on executive salaries to address income inequality and promote fair wages for all employees**. The system responded with the group title "Progressive Policy Advocates." Then, we activated AI to generate comments for the same set of strong-performing elements. The AI system analyses each of the three elements of values higher than 21 and, at the end, makes a general statement about the set of three elements. The third request for the AI system is to list and briefly describe attractive new or innovative products, services, experiences, or policies for this segment. The fourth and last request is to describe the messages that interest this segment. The following presents the dialog with the AI system for mind-set 1 of the two mind-set solution. This service is not activated for mind-set 2 as this mind-set does not have elements with values 21 or higher.

Mind-set 1 of 2—Top
Government Policies human-Based experiment
The following is a description of attributes from the total sample of a research study. Interested in: Implement a dynamic regulatory policy framework that can adapt to changing market conditions; introduce a cap on executive salaries to address income inequality and promote fair wages for all employees;
Create a label for this segment:
"Progressive policy advocates"
Describe this segment:
This segment of the total sample is likely to be socially conscious and value economic equity and fairness. They are likely to be in favor of implementing regulatory policies that are flexible and responsive to market changes, as well as in favor of measures to address income inequality, such as capping executive salaries. They are likely to prioritize fair wages for all employees and believe in the importance of reducing disparities in wealth and income. This segment may also be interested in addressing systemic issues that contribute to economic inequality and advocating for sustainable and ethical business practices
Describe the attractiveness of this segment as a target audience:

This segment of the total sample of the research study appears to be socially conscious and concerned with creating a more equitable and fair society. They are likely to be supportive of policies that promote economic equality and fair wages for all employees. They also value adaptability and regulation in the market to ensure that changing conditions are effectively addressed. As a target audience, this segment could be seen as highly attractive for companies or organizations looking to appeal to individuals who prioritize social responsibility and fairness in business practices. They may be more likely to support businesses that share their values and actively work toward creating a more just and equitable society
Explain why this segment might not be attractive as a target audience:
This segment might not be attractive as a target audience because they are likely to be more resistant to regulatory changes that could impact executive salaries and may not be open to policies that limit their earning potential. They may also be less supportive of initiatives aimed at addressing income inequality, as they may benefit from the current structure. As such, they may not be receptive to messages promoting fairness and regulation in the market. Additionally, this group may have a higher level of influence and power in decision-making processes, making them less likely to be swayed by external advocacy efforts
List what is missing or should be known about this segment, in question form:
1. What specific market conditions are expected to change, and how will the regulatory policy framework adapt to these changes? 2. How will the cap on executive salaries be determined and enforced? 3. How will the introduction of a salary cap impact the overall financial stability and performance of the companies in the segment? 4. What measures will be implemented to ensure fair wages for all employees beyond just addressing executive salaries? 5. How will the effectiveness of these policy changes be measured and evaluated over time?
List and briefly describe attractive new or innovative products, services, experiences, or policies for this segment:
1. Dynamic regulatory policy framework: This framework will allow for flexibility and adaptability in response to changing market conditions. It will ensure that regulations are in line with current needs and trends, helping to create a more stable and competitive business environment 2. Cap on executive salaries: By introducing a cap on executive salaries, income inequality can be addressed and fair wages can be promoted for all employees. This will help to create a more equitable workplace and distribute resources more evenly among all members of the organization Attractive new or innovative products, services, experiences, or policies for this segment could include: – A subscription-based model for services or products, offering customers the flexibility to pay for what they use on a regular basis – Implementing a remote work policy to allow employees to work from anywhere, promoting work-life balance and increasing productivity – Introducing sustainable and eco-friendly products or services to align with customer preferences and address environmental concerns – Incorporating artificial intelligence and automation into business processes to improve efficiency and streamline operations. – Creating a company culture focused on diversity and inclusion, promoting a positive work environment and attracting top talent
Which messages will interest this segment?

The segment of the total sample that is interested in implementing a dynamic regulatory policy framework and introducing a cap on executive salaries to address income inequality and promote fair wages for all employees will likely be interested in messages that focus on promoting social and economic justice, advocating for fairness in the workplace, and creating a more equitable society. Messages that highlight the importance of regulating market conditions to protect consumers and workers, ensuring that executive salaries are not disproportionately high compared to employee wages, and promoting a more level playing field for all individuals may resonate with this segment. Additionally, messages that emphasize the potential benefits of these policy changes in reducing income inequality, increasing economic stability, and fostering a more just and inclusive society may also be of interest to this segment

The following presents the dialog with the AI summarizer system for mind-set 1 of the three mind-set solution for the human-based experiment. This mind-set has only two elements of value 21 or higher that are **Implement a dynamic regulatory policy framework that can adapt to changing market conditions and Introduce a "regulatory impact assessment" requirement for all proposed policies.**

Mind-set 1 of 3—Top
The following is a description of attributes from the total sample of a research study. Interested in: Implement a dynamic regulatory policy framework that can adapt to changing market conditions, introduce a "regulatory impact assessment" requirement for all proposed policies;
Create a label for this segment:
– Adaptive regulatory policy – Regulatory impact assessment – Market conditions – Policy framework – Dynamic regulation – Segment label: Proactive regulation adaptation (PRA)
Describe this segment:
This segment of the total sample is likely made up of individuals who are interested in creating a regulatory policy framework that is flexible and responsive to changes in the market. They advocate for the implementation of a dynamic regulatory policy framework that can adapt to changing market conditions, showing a willingness to evolve policies as needed. Additionally, they support the introduction of a "regulatory impact assessment" requirement for all proposed policies, indicating a desire for thorough analysis and evaluation of the potential impacts of regulations before they are implemented. Overall, this segment values effective and efficient regulation that can support economic growth and stability
Describe the attractiveness of this segment as a target audience:
The sample shows a strong interest in creating a regulatory policy framework that is responsive to fluctuations in market conditions, demonstrating a forward-thinking approach to governance. The introduction of a "regulatory impact assessment" requirement highlights a commitment to thorough analysis and evaluation of proposed policies, indicating a dedication to informed decision-making. This segment exhibits a high level of engagement and commitment to ensuring that regulatory measures are effective, efficient, and tailored to suit the needs of the market. As a target audience, this segment is attractive for its proactive and strategic approach to regulation and its emphasis on evidence-based policy development. Their focus on adaptability and thorough evaluation suggests a willingness to engage in constructive dialogue and collaboration to achieve regulatory goals
Explain why this segment might not be attractive as a target audience:

This segment might not be an attractive target audience because they are likely to be advocates for more stringent regulatory measures and may be resistant to business-friendly policies. They may also have a vested interest in maintaining a strict regulatory environment and may not be receptive to proposals that prioritize flexibility and adaptability over strict regulations. Additionally, this segment may already have strong relationships with regulatory agencies and may be less open to engaging with new policy frameworks. Overall, targeting this segment may be challenging as they may not align with the desired goals of implementing a dynamic regulatory policy framework
List what is missing or should be known about this segment, in question form:
1. What specific market conditions necessitate a dynamic regulatory policy framework? 2. How will the dynamic regulatory policy framework be implemented and regularly assessed for effectiveness? 3. What criteria will be considered in the "regulatory impact assessment" requirement for proposed policies? 4. How will the impact assessment account for potential unintended consequences of regulatory policies? 5. What stakeholders will be involved in the assessment and decision-making process for regulatory policies? 6. How will the effectiveness of the regulatory policies be monitored and evaluated over time?
List and briefly describe attractive new or innovative products, services, experiences, or policies for this segment:
1. Implement a dynamic regulatory policy framework that can adapt to changing market conditions: This attribute suggests that the participants in the study value flexibility and agility in regulatory policies. They recognize the importance of responding promptly to shifts in the market to ensure sustainable growth and competitiveness 2. Introduce a "regulatory impact assessment" requirement for all proposed policies: This attribute indicates that the participants prioritize thorough analysis and evaluation of the potential effects of regulatory policies before implementation. They value evidence-based decision-making and seek to minimize any unintended consequences that may arise from new regulations Attractive new or innovative products, services, experiences, or policies for this segment may include: 1. Smart regulation technology platforms that can automate regulatory impact assessments and provide real-time data on market conditions to inform policy decisions 2. Collaborative regulatory networks that bring together industry stakeholders, government agencies, and consumer groups to co-create and evaluate regulatory policies 3. On-demand regulatory consultation services that provide personalized guidance and support to businesses navigating complex regulatory landscapes 4. Regulatory sandbox environments that allow for experimentation and piloting of new policies in a controlled setting before full implementation 5. Gamified compliance solutions that incentivize businesses to comply with regulations through rewards and recognition
Which messages will interest this segment?
1. "Stay ahead of the curve with our adaptable regulatory policy framework." 2. "Ensure the success of your proposed policies with our regulatory impact assessment tool." 3. "Navigate changing market conditions with confidence through our dynamic regulatory policy updates."

Similarly, we can obtain information about other mind-sets with elements of value 21 or higher for each experiment.

10.8 Conclusions

This chapter compares two experiments on government policies using the same set of pillars. Researchers and AI create the elements/answers for each pillar. The goal is to study whether the fact that AI is used to formulate the elements for each pillar in one of the experiments will affect the final results.

To evaluate both studies' performances, the Index of Divergent Thought (IDT) is used (Moskowitz et al. 2020). Experience shows that IDT values of 70 or so reflect stronger segmentations and more dissimilar clustering. In this study, the value of IDT for the AI-based experiment is 73, and for the human-based experiment, it is 60. The AI-generated elements more powerfully demonstrate differences among mind-sets.

Another interesting result is the number of strong-performing elements that will be the input for the AI summarizer. Summarizers tend to be wordy when they have a large set of high-scoring elements as input and very concise when dealing with a small set.

Summarizers, a unique tool in our arsenal, offer a wealth of recommendations when analyzing an input set. These insights are invaluable, as they bring to light perspectives that may not always be discernible to the human eye, thereby augmenting and enriching human arguments.

References

Asadullah, A., I. Faik, and A. Kankanhalli. 2018. "Digital Platforms: A Review and Future Directions." In *PACIS 2018 PROCEEDINGS*,.

Cai, Meng, and Ximing Yue. 2020. "The Redistributive Role of Government Social Security Transfers on Inequality in China." *China Economic Review* 62: 101512. https://linkinghub.elsevier.com/retrieve/pii/S1043951X20301097.

Cairney, Paul. 2021. "The Politics of Policy Design." *EURO Journal on Decision Processes* 9: 100002. https://linkinghub.elsevier.com/retrieve/pii/S2193943821001187.

Cairney, Paul. 2023. "The Politics of Policy Analysis: Theoretical Insights on Real World Problems." *Journal of European Public Policy* 30(9): 1820–38. https://www.tandfonline.com/doi/full/10.1080/13501763.2023.2221282.

Christainsen, G. B., and R. H. HAVEMAN. 1982. "Government Regulations and Their Impact on the Economy." *American Academy of Political and Social Sciences*.

Costas-Fernández, Julián, and Simón Lodato. 2022. "Inequality, Poverty and the Composition of Redistribution." *Social Choice and Welfare* 59(4): 925–67. https://link.springer.com/10.1007/s00355-022-01415-5.

Giancola, Marco, Simonetta D'Amico, and Massimiliano Palmiero. 2023. "Working Memory and Divergent Thinking: The Moderating Role of Field-Dependent-Independent Cognitive Style in Adolescence." *Behavioral Sciences* 13(5): 397. https://www.mdpi.com/2076-328X/13/5/397.

Gofman, A., and H. Moskowitz. 2012. *Rule Developing Experimentation: A Systematic Approach to Understand and Engineer the Consumer Mind*. eds. Alex Gofman (Deceased) and Howard R. Moskowitz. BENTHAM SCIENCE PUBLISHERS.

Graafland, Johan, and Lans Bovenberg. 2020. "Government Regulation, Business Leaders' Motivations and Environmental Performance of SMEs." *Journal of Environmental Planning and Management* 63(8): 1335–55. https://www.tandfonline.com/doi/full/10.1080/09640568.2019.1663159.

Hinterleitner, M., C. Knill, and Y. Steinebach. 2023. "The Growth of Policies, Rules, and Regulations: A Review of the Literature and Research Agenda." *Regulation & Governance* 18(6).

Howlett, M. 2021. "POLICY DESIGN WHAT, WHO, HOW, AND WHY." In *Handbook of Public Policy*,.

Kjaer, Poul F., and Antje Vetterlein. 2018. "Regulatory Governance: Rules, Resistance and Responsibility." *Contemporary Politics* 24(5): 497–506. https://www.tandfonline.com/doi/full/10.1080/13569775.2018.1452527.

Madore, Kevin P., Helen G. Jing, and Daniel L. Schacter. 2016. "Divergent Creative Thinking in Young and Older Adults: Extending the Effects of an Episodic Specificity Induction." *Memory & Cognition* 44(6): 974–88. http://link.springer.com/10.3758/s13421-016-0605-z.

Milutinovic, V., and J. Salom. 2016. *Mind Genomics: A Guide to Data-Driven Marketing Strategy (SpringerBriefs in Business)*. 1st Editio. eds. V Milutinovic and Salom.J. Springer.

Moskowitz, Howard R., Alex Gofman, Jacqueline Beckley, and Hollis Ashman. 2006. "Founding a new science: mind genomics." *Journal of Sensory Studies* 21(3): 266–307. https://onlinelibrary.wiley.com/doi/10.1111/j.1745-459X.2004.00066.x.

Moskowitz, H. R., Wren, J., & Papajorgji, P. (2020). Mind genomics and the law. LAP LAMBERT Academic Publishing.

Mucherino, A., Papajorgji, P. J., & Pardalos, P. M. (2009). Clustering by k-means. In Data Mining in Agriculture (pp. 47–82). Springer. https://doi.org/10.1007/978-0-387-88615-2_3

Mukherjee, Ishani, M. Kerem Coban, and Azad Singh Bali. 2021. "Policy Capacities and Effective Policy Design: A Review." *Policy Sciences* 54(2): 243–68. https://link.springer.com/10.1007/s11077-021-09420-8.

OpenAI. 2024. "ChatGPT [Large Language Model]." https://chat.openai.com.

Papajorgji, P., O. Ilollari, A. Civici, and H. Moskowitz. 2021. "A Mind Genomics-Based Cartography to Assess the Effects of the COVID19 Pandemic in the Tourism Industry." *WSEAS TRANSACTIONS ON ENVIRONMENT AND DEVELOPMENT* 17: 1021–29. https://wseas.com/journals/ead/2021/b905107-1444.pdf.

Patanakul, Peerasit, and Jeffrey K. Pinto. 2014. "Examining the Roles of Government Policy on Innovation." *The Journal of High Technology Management Research* 25(2): 97–107. https://linkinghub.elsevier.com/retrieve/pii/S1047831014000133.

Persson, Mikael. 2021. "From Opinions to Policies: Examining the Links between Citizens, Representatives, and Policy Change." *Electoral Studies* 74: 102413. https://linkinghub.elsevier.com/retrieve/pii/S0261379421001256.

Popiel, Pawel. 2022. "Digital Platforms as Policy Actors." In , 131–50. https://link.springer.com/10.1007/978-3-030-95220-4_7.

Röth, Leonce, and Hanna Schwander. 2021. "Greens in Government: The Distributive Policies of a Culturally Progressive Force." *West European Politics* 44(3): 661–89. https://www.tandfonline.com/doi/full/10.1080/01402382.2019.1702792.

Shleifer, A. 2005. "Understanding Regulation." *European Financial Management* 11(4): 439–45.

Thakral, Preston P., Amanda C. Yang, Donna Rose Addis, and Daniel L. Schacter. 2021. "Divergent Thinking and Constructing Future Events: Dissociating Old from New Ideas." *Memory* 29(6): 729–43. https://www.tandfonline.com/doi/full/10.1080/09658211.2021.1940205.

Todri, A., P. Papajorgji, and H. R. Moskowitz. 2024. "Investigating Gender-Based Responses in Social Psychology Experiments." *Journal of Cognitive Science* 25(1): 33–66.

Todri, A., P. Papajorgji, H.R. Moskowitz, and F. Scalera. 2020. "Perceptions Regarding Distance Learning in Higher Education, Smoothing the Transition." *Contemporary Educational Technology* 12(1).

Zdaniuk, Bozena. 2014. "Ordinary Least-Squares (OLS) Model." In *Encyclopedia of Quality of Life and Well-Being Research*, Dordrecht: Springer Netherlands, 4515–17. http://link.springer.com/10.1007/978-94-007-0753-5_2008.

Zemel, R. et al. 2019. "Consumers, and Dairy: Applying Artificial Intelligence, Mind Genomics, and Predictive Viewpoint Typing." In *Handbook of Modern Dairy Science and Technology*, IntechOpen.

Zhang, Mengke, Yan Huang, Yifan Jin, and Yuan Bao. 2023. "Government Regulation Strategy, Leading Firms' Innovation Strategy, and Following Firms Imitation Strategy: An Analysis Based on Evolutionary Game Theory" ed. Junyuan Yang. *PLOS ONE* 18(6): e0286730. https://dx.plos.org/10.1371/journal.pone.0286730.

Zhu, Jianhua et al. 2023. "Government Regulatory Policies for Digital Transformation in Small and Medium-Sized Manufacturing Enterprises: An Evolutionary Game Analysis." *Humanities and Social Sciences Communications* 10(1): 751. https://www.nature.com/articles/s41599-023-02250-4.

Correction to: The Mind of Everyday

Correction to:
P. Papajorgji, H. Moskowitz, *The Mind of Everyday*,
https://doi.org/10.1007/978-3-031-78078-3

The author name, Howard Moskowitz, was spelled incorrectly in the original version of the book in the front matter. This has been corrected now.

The updated version of this book can be found at
https://doi.org/10.1007/978-3-031-78078-3

P. Papajorgji, H. Moskowitz, *The Mind of Everyday*,
https://doi.org/10.1007/978-3-031-78078-3_11

Index

P. Papajorgji, H. Moskowitz, *The Mind of Everyday*,
https://doi.org/10.1007/978-3-031-78078-3

MIX
Papier aus verantwortungsvollen Quellen
Paper from responsible sources
FSC® C105338

If you have any concerns about our products,
you can contact us on
ProductSafety@springernature.com

In case Publisher is established outside the EU,
the EU authorized representative is:
Springer Nature Customer Service Center GmbH
Europaplatz 3, 69115 Heidelberg, Germany

Printed by Libri Plureos GmbH
in Hamburg, Germany